Springer Aerospace Technology

The *Springer Aerospace Technology* series is devoted to the technology of aircraft and spacecraft including design, construction, control and the science. The books present the fundamentals and applications in all fields related to aerospace engineering. The topics include aircraft, missiles, space vehicles, aircraft engines, propulsion units and related subjects.

More information about this series at http://www.springer.com/series/8613

Vereshchagin A. V. · Zatuchny D. A. ·
Sinitsyn V. A. · Sinitsyn E. A. ·
Shatrakov Y. G.

Signal Processing of Airborne Radar Stations

Plane Flight Control in Difficult Meteoconditions

Springer

Vereshchagin A. V.
St. Petersburg, Russia

Zatuchny D. A.
Moscow, Russia

Sinitsyn V. A.
St. Petersburg, Russia

Sinitsyn E. A.
St. Petersburg, Russia

Shatrakov Y. G.
St. Petersburg, Russia

ISSN 1869-1730 ISSN 1869-1749 (electronic)
Springer Aerospace Technology
ISBN 978-981-13-9987-9 ISBN 978-981-13-9988-6 (eBook)
https://doi.org/10.1007/978-981-13-9988-6

This Springer imprint is published by the registered company Springer Nature Singapore Pte Ltd.
The registered company address is: 152 Beach Road, #21-01/04 Gateway East, Singapore 189721, Singapore

Contents

Abbreviations

AA	Airborne avionics
AC	Amplitude characteristic
ACF	Autocorrelation function
ACM	Autocorrelation matrix
ADC	Analog to digital converter
AE	Antenna element
AFC	Automatic-frequency control
AFCh	Amplitude-frequency characteristic
ALC	Automatic level control
APC	Antenna system phase centre
APD	Amplitude-phase distribution
AR	Autoregression
ARMA	Autoregression and moving average
AS	Antenna system
ASC	Antenna system of coordinates
ASp	Air space
ATC	Air traffic control
AV	Air vehicle
AvC	Aviation complex
BCS	Bound coordinate system
CM	Centre of mass
CO	Coherent oscillator
computer	Computer
DP	Directional path
DS-P	Digital signal processing
DSP	Digital signal processor
EER	Effective echoing ratio
EMS	Elastic mode of a structure
EMW	Electro-magnetic wave
FFT	Fast Fourier transformation

FOS	Flight operation safety
FPGA	Field-programmable gate array
HM	Hydrometeor
HRF	High repetition frequency
ICAO	International civil aviation organization (international civil aviation organization)
IFA	Intermediate frequency amplifier
INS	Inertia navigation system
LRF	Low repetition frequency
LSL	Level of side lobes
LSM	Least square method
MA	Moving average
MD	Memory device
ML	Maximum likelihood
MO	Meteorological object
MRF	Medium repetition frequency
MUSIC	*Multiple signal classification* (multiple signal classification method)
OBC	Onboard computer
PAA	Phased antenna array
PCCU	Phase centre (position) control unit
PCU	Phase correction unit
PFC	Phase-frequency characteristic
PrS	Probing signal
PS	Phase shifter
PSD	Power spectral density
PSP	Programmable signal processor
R	Radar
RAM	Random access memory
RMS	Root mean square
RMSE	Root mean square error
ROM	Read-only memory
RP	Random process
RR	Radar reflectivity
RS	Radar station
SINR	Signal–interference–noise ratio
SMT	Selection of moving targets
SNR	Signal–noise ratio
software	Software
specification	Specification
TI	Trajectory irregularities
UHF	Ultra-high frequency
US	Underlying surface
WS	Wind shear
WSAA	Waveguide slot array antenna

Chapter 1
The Detection and Assessment of Meteorological Object Parameters by Airborne RS

1.1 Flight of Air Vehicles in Wind and Atmospheric Turbulence

One of the most commonly occurring adverse environmental conditions in the practice of AV flights leading to consequences, dangerous on the influences on AV are various perturbations of the atmosphere (Table 1.1) [1]. Owing to the growth of intensity of flights and decrease of requirements to admissible weather conditions,the probability of AV entrance to such zones has considerably grown recently.

1.1.1 The Spatial Wind Field in the Earth's Boundary Layer

For the assessment of the influence of wind on AV flight, we will consider a wind speed vector $\vec{V}$ in some point of space and its projections V_x, V_y, V_z on the axis of the Earth-based coordinate system of (Fig. 1.1). In the turbulent atmosphere, air flow passes through irregular changes on all axes, owing to what pulsations appear. At the same time the total wind speed on each of the axes will consist of an average value $\bar{V}$ and a random deviation V'

$$V_x = \bar{V}_x + V'_x \quad V_y = \bar{V}_y + V'_y, \quad V_z = \bar{V}_z + V'_z,$$

where $\bar{V}_x$, $\bar{V}_y$, $\bar{V}_z$ and V'_x, V'_y, V'_z—average values of wind speed and its deviation from average values in the directions of axes of coordinates.

Thus, changes in the characteristics of the spatial field of wind speed on three axes OX, OY, and OZ can be described by 18 components

$$\frac{\partial \bar{V}_x}{\partial x}, \frac{\partial \bar{V}_x}{\partial y}, \frac{\partial \bar{V}_x}{\partial z}; \quad \frac{\partial V'_x}{\partial x}, \frac{\partial V'_x}{\partial y}, \frac{\partial V'_x}{\partial z};$$

Vereshchagin A. V. et al., *Signal Processing of Airborne Radar Stations*,
Springer Aerospace Technology, https://doi.org/10.1007/978-981-13-9988-6_1

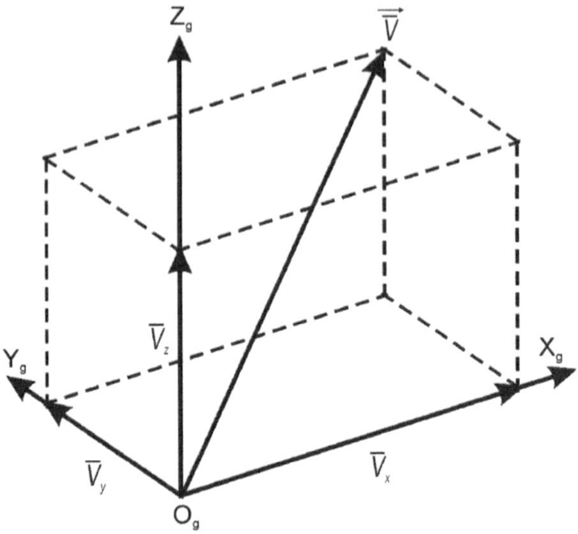

Fig. 1.1 The components of wind speeds concerning the Earth-based coordinate system

Table 1.1 The list of the atmospheric perturbations potentially dangerous for AV flights

Item no.	Parameter	Parameter value			
		Weak	Moderate	Heavy	Very heavy
1	Vertical wind shear at the height of 30 m, m/s	0–2.0	2.1–4.0	4.1–6.0	>6
2	Horizontal wind shear at the distance of 600 m, m/s	0–2.0	2.1–4.0	4.1–6.0	>6
3	RMSD of velocity fluctuations, m/s	0–1.5	1.5–3.0	3.0–4.5	>4.5
4	Overload, g	0–0.2	0.2–0.4	0.4–0.6	>0.6
5	Speed of dissipation of turbulent kinetic energy, m^2/s^3	0–0.001	0.001–0.01	0.01–0.04	>0.04
6	Influence on AV control	Insignificant	Significant	Considerable	Dangerous

$$\begin{aligned} &\frac{\partial \bar{V}_x}{\partial x}, \frac{\partial \bar{V}_x}{\partial y}, \frac{\partial \bar{V}_x}{\partial z}; \quad \frac{\partial V_x'}{\partial x}, \frac{\partial V_x'}{\partial y}, \frac{\partial V_x'}{\partial z}; \\ &\frac{\partial \bar{V}_z}{\alpha}, \frac{\partial \bar{V}_z}{\partial y}, \frac{\partial \bar{V}_z}{\partial z}; \quad \frac{\partial V_z'}{a}, \frac{\partial V_z'}{\partial y}, \frac{\partial V_z'}{a}. \end{aligned} \tag{1.1}$$

The spatial changes of the average values of wind speed characterize the wind shears in an air layer on the corresponding axes, and small-scale random changes of speed (with a line)—the turbulence degree.

1.1.2 Wind Shear

The wind shear (WS), according to the recommendations of ICAO, is equal to the vector difference of wind speeds in two distanced points of space to the distance between them [2, 3]. WS represents a sharp change of wind speed and/or the direction at which an AV is sharply displaced in relation to the planned trajectory and demands additional actions of a crew of an AV control [3, 4].

The influence of WS on an AV flight is that in the conditions of sharp change of wind speed and the direction the conditions of the flow of an AV with the approaching stream (air speed, angles of attack, sliding) change. At the same time, owing to inertia of an AV mass and delay in change of an engine's operating mode, travelling speed temporarily remains. The sharp change of air speed causes corresponding change of the aerodynamic (lifting) force directly proportional to a square of air speed that leads to violation of balance of forces and the moments having an effect on an AV [3–5]. The transition process on restoration of the mode of flight demands certain time for which an AV significantly deviates from the set trajectory depending on an WS size, time of its influence and delay on its counteracting. The special difficulty for piloting is presented by cases in which the change of a WS sign to an opposite one is observed along a trajectory of an AV flight. In these cases, a delay of the actions of a crew for compensation of the previous influence can unsuccessfully develop with new influence. For example, reduction of engine's draught at the simultaneous falling of air speed at the expense of WS can lead to dangerous total influence on an AV.

The WS vector $\vec{V}_в$ can be characterized by the module and angle of rotation

$$|\vec{V}_в| = \partial\bar{V}/\partial r \approx \Delta\bar{V}/\Delta r, \tag{1.2}$$

$$\varphi_в = \angle(\vec{V}_1\vec{V}_2), \tag{1.3}$$

where

$\Delta\bar{V} = \left|\vec{V}_2 - \vec{V}_1\right|$ the module of difference of wind speed vectors in two points;

$\angle\left(\vec{V}_1\vec{V}_2\right)$ the angle between wind speed vectors in two points;

Δr the distance between points (usually is 30 m in height or 600 m in distance [6, 7]).

The definition of WS is carried out by the results of measurements of wind speed, filtered from small-scale turbulent pulsations.

On the other hand, according to (1.1), WS can be estimated by means of the corresponding projections of speed. From the components listed above, the greatest danger to AV is constituted by the components $\partial\bar{V}_x/\partial x$ and $\partial\bar{V}_x/\partial z$ [3]. The changes of $\partial\bar{V}_x/\partial y$ and $\partial\bar{V}_z/\partial x$ in the layer Δz are rather small, and their influence on AV flight can be neglected, as, usually an AV passes in a horizontal direction the distance by 1–2 orders of magnitude more than in vertical direction. However, in some cases (for example, supercellular thunderstorms), there can be microvortexes when the values of $\partial \bar{V}_z / \partial z$ can become considerable and dangerous (more than 10 m/s at 30 m of height) but their probability is small and does not exceed 10^{-8} [4].

The measure of danger of WS is the so-called F-factor (or the Wind shear hazard index) offered by Bowles [8, 9] within the works carried out by NASA together with Rockwell Collins Avionics (USA). The F-factor characterizes the change of total energy of AV owing to the influence of wind shear and is defined by the expression

$$F = \dot{V}_x / g - W_Z / W_x, \tag{1.4}$$

where

$\dot{V}_x$	the speed of change of a horizontal projection of wind speed;
G	downward acceleration;
W_x and W_z	horizontal and vertical projections of an AV air speed.

The positive value of the F-factor corresponds to reduction of AV total energy, v. The reduction of AV total energy happens either due to the reduction of its potential energy (at the descending of AV, $W_z < 0$) or due to the reduction of its kinetic energy owing to strengthening of a front wind, $\dot{V}_x > 0$.

The speed of change of a horizontal projection of wind speed at the same time is the function of meteorological characteristics of wind and parameters of an AV trajectory

$$\dot{V}_x = \frac{\partial V_x}{\partial x}\dot{x} + \frac{\partial V_x}{\partial z}\dot{z} + \frac{\partial V_x}{\partial t}. \tag{1.5}$$

The first summand in (1.5) is the result of the product of horizontal WS along an AV trajectory $\partial V_x / \partial x$ and an AV ground speed $\dot{x}$. The second summand represents the product vertical shear of longitudinal wind $\partial V_x / \partial z$ and the speed of descending of an AV $\dot{z}$. The third summand characterizes the rapidness of longitudinal wind speed change over time.

The expressions (1.4) and (1.5) characterize the WS influence only on instantaneous values of an AV total energy and its air speed. It is necessary to average the F-factor at a certain distance along a flight trajectory for using it as the index of danger. Based on the results of full-scale measurements using laboratory aircraft and mathematical modeling, the FAA established a distance of 1 km for averaging. At the same time, the values $\bar{F} > 0, 1$ are defined as dangerous and the threshold of development of an alarm signal corresponds to the values $\bar{F} \geq 0, 13$.

1.1.3 Atmospheric Turbulence

The main reason for the atmospheric turbulence are the contrasts in the field of wind and the field of temperature generated by various processes: big vertical gradients of wind in a lower layer, deformation of air currents by mountains, unequal heating of various sites of underlying surface, cloud formation processes, interaction of air masses with different properties. Air whirling motion in the atmosphere is usually characterized by the existence of vortexes of various sizes moving with various speeds in a common (average) flow.

Turbulent pulsations, especially pulsations of vertical wind speed (vertical gusts) cause sharp AV movements in the vertical plane—the "bumpiness" characterized by the emergence of sign-variable accelerations, linear fluctuations of the AV centre of gravity and angular fluctuations of its case concerning the centre of gravity. At the same time, the height, course, speed and mode of AV flight change suddenly and in considerable limits, AV stability and controllability worsen. Besides, "bumpiness" causes additional loads of the structural elements increasing wear of the AV separate units. The extent of influence of atmospheric turbulence on an AV flight depends on the AV sizes and the disturbed zone through which it passes, on speed and direction of their mutual movement defining rapidity of air speed change, an angle of attack and sliding of AV.

It was experimentally found [4, 10] that "bumpiness" in clouds at all levels in the atmosphere occurs considerably more often than in a clear sky. At the same time, its repeatability in clouds of various forms is not identical and depends on the physical causes of occurrence of cloud cover of this or that type. Most often, intensive "bumpiness" occurs in cumuliform clouds. At the same time, turbulent gusts (vortexes) have rather small sizes (in cumulus clouds—to several tens of metres, and in Cumulonimbus clouds—up to 1000 m), comparable with AV sizes.

Let us consider projections of turbulent motions to axes of the Earth-based coordinate system (Fig. 1.1). Wind gusts along the axes *OX* and *OY*, on average, are insignificant, are 0.2–0.5 m/s and, as a rule, can be not considered [11] in calculations though side winds that cause a vigorous rolling motion of an AV owing to action of the moment of roll stability [12]. The exception is made by the conditions created by small-scale vortex rotor circulation (here the values of $\partial V_x'/\partial z$ and $\partial V_z'/\partial z$ often exceed 10 m/s). Turbulent vertical gusts, besides the emergence of "bumpiness" are also dangerous because they can lead an AV to angles of attack where stability on an angle of attack is decreased or lost.

The field of speeds of the small-scale air traffics caused by turbulence as a first approximation can be described within Kolmogorov–Obukhov theory of uniform and isotropic turbulence of [13]. Here, it should be noted that the border between the large-scale component caused by stratification of the wind field and also by the existence of the ascending and descending convective streams, and a small-scale component of movements of air masses is very conditional. The spatial radius of correlation of not uniformity of wind streams' speed or external scale of turbulence of L0 can be such border. In clouds, the value L0 can change from hundreds of metres to kilometre units [14].

According to Kolmogorov–Obukhov theory, in the inertial interval of scales (l0, L0) where l0 is the internal scale of turbulence, the velocity field of turbulent motion can be considered as "frozen" (Taylor's hypothesis) for all the time of a flight. It is caused by the fact that owing to the high own flight velocity in comparison with speeds of turbulent motion, an AV air flies the distance throughout which the correlation connection between these velocities is rather strong, so quickly that during a flight of this distance, the wind velocity field does not change significantly. It means that for uniform isotropic turbulence the distribution of energy will be invariant in relation to the spatial section of a turbulent stream in the direction of the AV movement and temporal section in any of the points of its trajectory.

The key indicator of atmospheric turbulence danger is overload n [10, 15], resulting from the AV "bumpiness". The value of an overload depends, both on characteristics of turbulent MO and on AV design parameters. On the other hand, the spatial field of turbulent gusts can be described by structural functions of small-scale changes of velocity components of the air environment. In particular, the structural function of the ith speed component is

$$D_i(\vec{r}) = \left\langle \left[V_i'(\vec{R} + \vec{r}) - V_i'(\vec{R}) \right]^2 \right\rangle,$$

where $\vec{r}$—the vector between any two points.

The value $D_i(\vec{r})$ is defined by the distance between points r and the speed of dissipation of the turbulent kinetic energy ε, characterizing the intensity of turbulence and can be presented by the expression

$$D_i(\vec{r}) = C_i^2 \varepsilon^{2/3} r^{2/3},$$

where C_i^2—the structural constant.

Let us note that the constant C_l^2 for a longitudinal (in relation to the direction of the vector $\vec{r}$) component of the speed and the constant C_t^2 for a transverse component of the speed are connected by the dependence

$$C_{tt}^2 = 4/3 C_{ll}^2.$$

The value l_0 of the internal scale of turbulence is defined by the value ε and kinematic viscosity v

$$l_0 = 5\pi \left(v^3/\varepsilon \right)^{1/4}$$

and has the order of magnitude of millimetres or centimetres within all thickness of the troposphere.

Summing up the result, it is possible to note the following features:

1. The danger of heavy WS consists of sharp change of AV air speed that leads to an essential deviation of AV from the set trajectory. The greatest danger to AV is constituted by longitudinal and vertical shears of longitudinal air speed, especially at stages of prelanding manoeuvering and AV landing approach.
2. Intensive turbulent pulsations of wind speed cause AV bumpiness. At the same time, the height, course, speed and mode of AV flight change suddenly and in considerable limits; AV stability and controllability worsen. The most intensive bumpiness is caused by the turbulent gusts comparable with AV sizes.

1.2 The Detection of Wind Shears, Dangerous for Flights and Zones of Intensive Turbulence with the Use of Airborne RS

The existing high correlation between areas of heavy convective traffics and condensation of water vapour (with cloud cover formation and precipitation) (the Annex 1) allows to use a small-drop atmosphere phase as a tracer of movements of air flows. The main facility for the detection of dangerous zones of intensive air traffics in cloudy MO is the RS of 3-centimetre range providing the considerable range of observation.

At radiation by EMW of 3-centimetre range, MO represents a spatially distributed radar R target consisting of a set of a large number of accidentally located and independently moving elementary reflectors—hydrometeors (HM) [16]. In this case, the problem of detection of MO and assessment of the degree of its danger comes down, in essence, to a problem of detection of an R signal reflected by MO against the background of jamming and noise and also assessment (measurement) of its DS parameters.

Let us break all MO volumes into the separate allowed volumes $V_u = V_u[\delta r, \delta\alpha, \delta\beta]$ (Fig. 1.2) whose sizes are determined by the RS resolution by distance, azimuth and a place angle—respectively δr, $\delta\alpha$, $\delta\beta$.

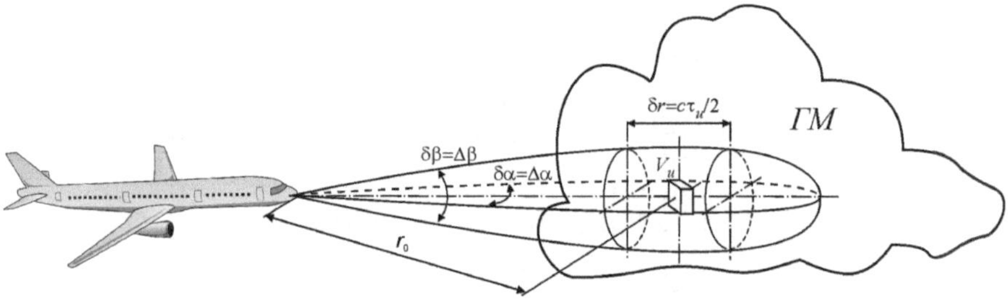

Fig. 1.2 The spatial position of the allowed volume of HM

Neglecting multiscattering and also absorption of EMW in MO volume, the total signal reflected by any allowed volume at an input of the RS receiver, it is possible to present superpositions of the components caused by each elementary reflector within the limits of *Vi* in the form

$$s(t) = \sum_{V_u} s_n(t), \tag{1.6}$$

where

$$s_n(t) = A_n(t)\exp[-j2\pi f_0 t + j\varphi_0] \tag{1.7}$$

the signal reflected by the *n*th HM entering the V_i volume;

$A_n(t)$ the amplitude of the signal reflected by the *n*th HM;
f_0 the carrier frequency of the RS probing signal (PS);
φ_0 the initial phase of the RS PS.

It should be noted that the signal of any HM is much less than the total signal reflected by *Vi* volume, i.e. it is supposed that separate large particles ("bright points") having the high effective echoing ratio (EER) and are not a part of MO.

Let us assume that each elementary reflector entering V_i volume moves with the speed $\vec{V}_n$. The difference of velocities of separate HM can be caused by a wind velocity gradient (shear) in the radial direction, turbulence, different influence of gravity on reflectors with a different weight. Under the influence of wind, the reflectors enter the allowed volume participate in two types of movement:

- large-scale, causing regular movement of all reflectors of this allowed volume;
- small-scale, representing random movements of reflectors along the centre of the allowed volume.

Thus, the velocity of any reflector contains two components and is defined by the vector expression

$$\vec{V}_n = \vec{\bar{V}} + \vec{V}_{\Delta n} \tag{1.8}$$

where

$\vec{V}_n$ the *n*th reflector speed vector as a part of the allowed volume;
$\vec{\bar{V}}$ the speed vector of the large-scale movement of reflectors;
$\vec{V}_{\Delta n}$ the speed vector of the small-scale movement of the *n*th reflector.

The value of a radial projection of the velocity of the *n*th reflector taking into account (3) can be presented in the form of the expression

$$V_n = \bar{V} + V_{\Delta n}, \tag{1.9}$$

where

V_n the radial projection of the speed of the *n*th reflector;
$\bar{V}$ the radial projection of the speed of the large-scale movement;
$V_{\Delta n}$ the radial projection of the speed of the small-scale movement of the *n*th reflector

If we introduce the *S(V)* function determining the distribution (spectrum) of radial speeds of reflectors for this allowed volume, then the maximum of *S(V)* corresponds to some average speed $\bar{V}$ of the movement of all volume V_i which can be accepted as the assessment of a radial component of speed of large-scale movement.

The dispersion of speeds of separate HM is relatively $\bar{V}$ characterized by the root mean square (RMS) width of the spectrum of radial speeds ΔV, which can be used for the assessment of the intensity of small-scale movements of reflectors.

Relocation of HM leads to the emergence in the reflected signal $s_n(t)$ of Doppler frequency shift

$$f_{\partial n}(t) = \pm 2 \frac{\partial_n}{\partial t} f_0 = \pm \frac{2V_n(t)}{c} f_0, \qquad (1.10)$$

where

r_n the range from an RS to the reflector;
c the velocity of distribution of EMW in space.

The sign of Doppler frequency shift is defined by the direction of the movement of the reflector in relation to RS.

Thus, the signal (1.6) reflected by a set of independently moving HM located in V_i volume will contain the spectrum of frequencies corresponding to a spectrum of the radial components of velocities of different elementary reflectors; and for any number of HMs in the allowed volume, the spectrum of velocities of particle movement is unambiguously connected with the Doppler spectrum (DS) of the radio signals reflected by MO [17]

$$S(V)dV = S(f_\partial)df_\partial. \qquad (1.11)$$

The mutual movement of reflectors leads to small widening of a signal spectrum in comparison with the carrier frequency. Therefore the signal reflected by the allowed volume will represent a narrow-band random process which in a time domain can be described by the expression

$$s(t) = A(t)\exp[-j2\pi f_0 t + j\varphi_0 + j\varphi(t)]. \qquad (1.12)$$

The amplitude A(t) and the phase φ(t) in (1.12) are random values as distribution of HM in space and in time changes in a random way, therefore, the signal reflected by the allowed MO volume fluctuates. Movements of reflectors along the directional pattern (DP) of the RS antenna system (AS) owing to wind, turbulence, scanning of

DP, movement of the RS carrier or other factors can be the reasons of fluctuations. The form and duration of fluctuations depend on the form and duration of the probing signal (PS), the DP form and also laws of the movement of the RS carrier, scanning of DP or movement of reflectors.

Detection of the R signal reflected by the allowed MO volume is a special case of the statistical theory task of testing hypothesis [18]. According to this theory, in the course of detection of an MO signal against the background of interferences and intra reception noise in each allowed volume the formation of the likelihood ratio or its unambiguous functionality [19], its threshold processing is carried out, and the decision on absence ("0") or existence ("1") of potentially dangerous object is made. The quality of detection of the signal reflected by MO can be estimated on the basis of the analysis of the operating characteristics of the RS receiver. The criteria of efficiency of detection are the probabilities of the correct detection p_D and false alarm p_F, representing the elements of a matrix of the probability of the second order [19]

$$\underline{\mathbf{p}} = \begin{bmatrix} p_{00} & p_{01} \\ p_{10} & p_{11} \end{bmatrix},$$

where

$p_{10} = p_F$	the probability of false alarm;
$p_{11} = p_D$	the probability of the correct detection of an MO signal;
p_{00}	the probability of the correct nondetection of an MO signal;
p_{01}	the probability of the admission of an MO signal.

Let us note that for the specified probabilities the following condition is satisfied

$$p_{00} + p_{10} + p_{01} + p_{11} = 1.$$

For the assessment of probabilistic criteria of efficiency of processing, it is necessary to define statistical characteristics of the signal (1.12) reflected by the allowed MO volume, in particular, the one-dimensional law of the distribution of the amplitude A. This distribution is connected with distribution of specific EER on the MO volume and is generally described by model of m-distribution of *M. Nakagami*) [19, 20]

$$w(A) = K_m A^{2m-1} \exp\left(-mA^2\right), \tag{1.13}$$

where $K_m = 2m^m / \Gamma(m)$—the normalizing constant.

Usually within the allowed volume, there is a large number of reflectors, the frequency of fluctuations is rather high and the area filled with reflectors is larger than V_i; thanks to what at the movement of the allowed volume along a beam at sounding a part of reflectors enters it, and a part that exits it, creating a combination of perturbations. In [21–23], it is shown that under such circumstances on the basis of the central limit theorem of probabilities, the considered random process can be considered to be Gaussian in the narrow sense, i.e. distribution densities of probabilities of any order are Gaussian. In this case, in the distribution (1.13), the value of the parameter $m = 1$ and Nakagami distribution turns into Rayleigh distribution

$$w(A) = 2A\exp\left(-A^2\right).$$

The procedure of detection of the signal reflected by the allowed MO volume comes down to the detection of the Rayleigh signal against the background of the fluctuating noise [19] by comparison of the useful signal–noise power ratio (SNR) q at the output of the RS receiver with the threshold value [24]

$$q_n = (\ln p_F / \ln p_D) - 1. \tag{1.14}$$

Considering strict requirements on flight operation safety in dangerous zones [25], the quality of MO detection should be characterized by the following indicators: $p_D \geq 0,9$ and $p_F \leq 10^{-4}$ [26, 27]. According to (1.14) in this case $q_n > 64$ (18 dB), and taking into account the coefficient of losses at processing 4–7 dB of SNR should be in the limits of 22–25 dB.

As the operational characteristics of the RS receiver representing dependences of probability of the correct detection of p_D from SNR q at the fixed values of probability of false alarm p_F are monotonously increasing functions, then instead of consideration of probabilistic criteria of efficiency of R information processing, it is possible to pass on to the analysis of the corresponding power criteria:

- SNR at the input of the threshold detection device;
- the improvement coefficient of SNR.

The SNR value at the input of the threshold device allows defining the MO observability and possibility of its detection as knowing the SNR value it is possible to calculate probabilities of the correct detection and false alarm for the developed detection algorithm. Besides, this criterion allows defining basic expediency of processing in the developed interference-target situation.

The improvement coefficient of SNR is defined by the expression [28]

$$K_{\text{у}} = 10\lg\frac{q_{\text{вых}}}{q_{\text{вх}}} = 10\lg\frac{(P_c/P_{ш})_{вых}}{(P_c/P_{ш})_{вх}} = 10\lg\frac{P_{c\,вых}}{P_{c\,вх}} + 10\lg\frac{P_{ш\,вх}}{P_{ш\,вых}} = K_c K_{ш} \tag{1.15}$$
,

where

P_c the power of the signal reflected by the allowed MO volume;
P_{III} the power of the fluctuating noise;
K_c the coefficient of strengthening of a useful signal;
$K_{ш}$ the coefficient of noise suppression.

The improvement coefficient (1.15) at the same time reflects both the extent of noise suppression and the extent of amplification of the useful signal reflected by MO.

Realization of only threshold processing by the detector of the average power of the reflected signals within the allowed volume is characteristic of the existing incoherent airborne meteo RS. However, owing to biunique communication (1.11), the DS of the reflected signals and the spectrum of radial MO speeds on the basis of the parameter assessment of the Doppler spectrum for each allowed volume of $V_{i,}$ it is possible to calculate the corresponding meteorological characteristics—the values of $\bar{V}$ and ΔV. In particular, three first moments of the Doppler spectrum can be used for the assessment of speed parameters of air masses movement $S(f_\partial)$ [29]:

(a) the zero moment of the Doppler spectrum representing the average power of the signals reflected by the allowed MO volume,

$$M_0[S(f_\partial)] = \bar{P}_c = \int S(f_\partial)df_\partial; \tag{1.16}$$

(b) the first moment of the spectrum equal to the average Doppler frequency proportional to the average radial speed of the allowed volume in general,

$$M_1[S(f_\partial)] = \bar{f_\partial} = \frac{1}{\bar{P}_c}\int f_\partial S(f_\partial)df_\partial = \frac{2\bar{V}}{\lambda}, \tag{1.17}$$

where $\lambda = c/f_0$—the RS PS wavelength;

(c) the second, central, moment equal to the DS dispersion and defining the dispersion ΔV^2 of speeds of reflectors within the allowed volume

$$M_2[S(f_\partial)] = (\Delta f_\partial)^2 = \frac{1}{\bar{P}_c}\int \left(f_\partial - \bar{f_\partial}\right)S(f_\partial)df_\partial = \frac{4\Delta V^2}{\lambda^2}. \tag{1.18}$$

The presence of the gradient of the wind speed (WS) $\bar{V}$ leads to the fact that the values of average radial wind speed in the spaced allowed volumes will be different. Therefore, for the determination of the WS value in a certain direction, it is necessary to estimate the difference of average Doppler frequencies in the allowed volumes spaced on some distance Δr. Then the wind velocity gradient (or WV value) (1.2) is equal to

$$\left|\vec{V}_в\right| = \frac{\partial\bar{V}}{\partial r} \approx \frac{\Delta\bar{V}}{\Delta r} = \frac{\left|\bar{V}_2 - \bar{V}_1\right|}{\Delta r} = \frac{\lambda}{2\Delta r}\left|\bar{f}_{\partial 2} - \bar{f}_{\partial 1}\right|. \tag{1.19}$$

The parameters of small-scale movements of the reflectors caused, first of all, by atmospheric turbulence can be determined by the analysis of dispersion (the RMS width) of DS in each allowed volume. For a case of uniform local and isotropic turbulence whose spectral density is described by Kolmogorov–Obukhov law; the root mean square width of a spectrum as it is noted in [16, 30], significantly depends on the scale of turbulence L

$$\Delta f_t^2 = A_t \lambda^{-2} C_t \varepsilon^{2/3} L^{2/3},$$

where A_t, C_t—the constants.

Owing to uniformity and isotropy of turbulence at scales L comparable with linear sizes of the allowed volume and also owing to Taylor's hypothesis of "frozen" turbulence [31], the DS of an MO signal caused by turbulent motion will be symmetric in relation to $\bar{f}_\partial$, corresponding to the average speed of the allowed volume.

The considered above SNR value also determines the potential accuracy of the assessment of not power parameters of the reflected signal which include the average frequency $\bar{f}_\partial$ and the width Δf_∂ of its DS. In particular, the accuracy of the assessment is characterized by the RMS value of an estimation error [32]

$$\sigma_\xi = \sqrt{D(\hat{\xi}\,|\xi) + b^2(\hat{\xi}\,|\xi)},$$

where

σ_ξ—the RMS error of the assessment

$\hat{\xi}$— of not power parameter of a signal from its true value ξ;

$D(\hat{\xi}|\xi) = M\left\{[\hat{\xi} - M\{\hat{\xi}\}]^2\right\}$ and $b(\hat{\xi}|\xi) = M\{\hat{\xi} - \xi\}$—and the dispersion and shift of assessment of not power parameter of a signal, respectively.

At an ideal straight-line characteristic of the measuring instrument (discriminator), the assessment of not power parameter is not displaced [19], and the accuracy is completely defined by the value of dispersion of the assessment whose minimum possible value is set by the Cramer–Rao inequality [24]

$$D(\hat{\xi}|\xi) = -\left[M\left\{\frac{\partial^2}{\partial \xi^2} \ln \Lambda(y|\xi)\right\}\right]^{-1},$$

where $\Lambda(y|\xi)$—the conditional likelihood ratio.

For a signal with the random amplitude A the initial phase φ_0 the expression for the Cramer–Rao bound of the assessment of Doppler frequency shift of an MO signal is brought to the following [19, 33]

$$D(f_\partial) = -\left[q^2 \frac{\partial^2}{\partial f_\partial^2} \rho(0, f_\partial)\Big|_{f_\partial=0}\right]^{-1} = \frac{1}{q^2 \tau^2_{экв}},$$

where $\rho(0, f_\partial)$—the section of two-dimensional-rated time–frequency autocorrelated function (ACF) [33, p. 202];

the equivalent duration of an R signal;

$\tau_{экв} = \sqrt{-\frac{\partial^2}{\partial f_\partial^2}\rho(0, f_\partial)\Big|_{f_\partial=0}}$ the equivalent duration of an R signal.

To provide the set accuracy of the assessment of Doppler frequency shift of an MO signal $D(f_\partial)$, it is necessary to have at the input of the processing device the SNR value

$$q_n = \left[1 - 2\rho_K k_Д \left|\hat{f}_\partial - f_\partial\right| + \left(k_Д \left|\hat{f}_\partial - f_\partial\right|\right)^2\right] \Big/ \left[k_Д^2 D\left(\hat{f}_\partial, f_\partial\right)\right], \tag{1.20}$$

where

ρ_k the coefficient of the interchannel correlation of signals;
$k_Д$ the slope of the discriminatory characteristic of the measuring instrument.

At the influence of the fluctuating noise $\rho_k = 1$ and for the not displaced assessment, the expression (1.20) is transformed to the following

$$q_n \approx \Delta F(1 - f_\partial/\Delta F)/\delta(f_\partial)$$

where ΔF—the measuring instrument bandwidth width (defined by the width of a spectrum of an MO signal depending on the range of possible pulsations of wind speed, the pattern width and the scanning law).

For a standard airborne RS at $\Delta V_{\mu\alpha\xi} = 10\,\text{m/s}$ [34] and the required value $\delta(f_\partial) = 1\,\text{m/s}$ [35], it is necessary to provide $q_n \geq 600$ (that is not less than 27 dB). Besides, at the determination of the threshold ratio, it is also necessary to consider the losses in a path of processing [36] arising because own noise of the RS receiver at the output of the processing device becomes correlated. As it is shown in [37], the losses, in this case, are no more than 2.5 dB.

The main statistical parameters characterizing the efficiency of the assessment of danger of WS and turbulence on the basis of the analysis of the moments of DS of the reflected signals are the probability of the correct assessment p'_D and the probability of false alarm p'_F, which are completely defined by the accuracy of estimates of the DS corresponding parameters.

As the estimates of the average speed and width of a spectrum of speeds of particles in the allowed volume of MO can be presented in the form of the sum of the corresponding average values (corresponding to true values of parameters) and the aligned random processes—noises of measurements, the probability of the correct assessment p'_D will be equal to the probability of not excess by an RMS error of the assessment of the corresponding parameter of permissible value $\sigma_{норм}$

$$p_D^{св} = p\left(\left|\sigma_{\overline{V}}\right| \le \sigma_{норм}\right) = \int_{-\sigma_{норм}}^{\sigma_{норм}} w_1(\sigma_{\overline{V}}) d\sigma_{\overline{V}}, \quad (1.21)$$

$$p_D^{m} = p\left(\left|\sigma_{\Delta V}\right| \le \sigma_{норм}\right) = \int_{-\sigma_{норм}}^{\sigma_{норм}} w_1(\sigma_{\Delta V}) d\sigma_{\Delta V}, \quad (1.22)$$

where

$p_D^{св}$	is the probability of the correct assessment of dangerous wind shear;
p_D^{m}	is the probability of the correct assessment of dangerous turbulence;
$w_1(\sigma_{\bar{V}})$ and $w_1(\sigma_{\Delta V})$	are one-dimensional densities of the probability of errors of the assessment of the average speed and RMS width of a spectrum of speeds of HM.

The probabilities of false alarm p_F at the detection of dangerous meteophenomena can be determined by the expressions

$$p_F^{св} = p\left(\sigma_{\overline{V}} > \sigma_{норм}\right) = \int_{\sigma_{норм}}^{\infty} w_1(\sigma_{\overline{V}}) d\sigma_{\overline{V}},$$

$$p_F^{m} = p\left(\sigma_{\Delta V} > \sigma_{норм}\right) = \int_{\sigma_{норм}}^{\infty} w_1(\sigma_{\Delta V}) d\sigma_{\Delta V}$$

The own movement of the RS carrier leads to additional movement of the MO disseminating particles in relation to an RS antenna system phase centre (ASFC) and, therefore, to the emergence in DS of the reflected signals of the additional components, distorting the valid spectrum of radial speeds of reflectors. This circumstance causes the need for consideration and compensation of own movement of the RS carrier at the assessment of the DS parameters of signals [38].

Existence of a radial component of the RS carrier velocity leads, first of all, to the shear (Fig. 1.3) of the whole DS of the reflected signals (without distortion of a form) along an axis of frequencies by the value

$$\bar{f}_{дв} = 2W_r/\lambda, \quad (1.23)$$

where

$W_r = W\cos\theta_0$	the radial component of the speed of the RS carrier;

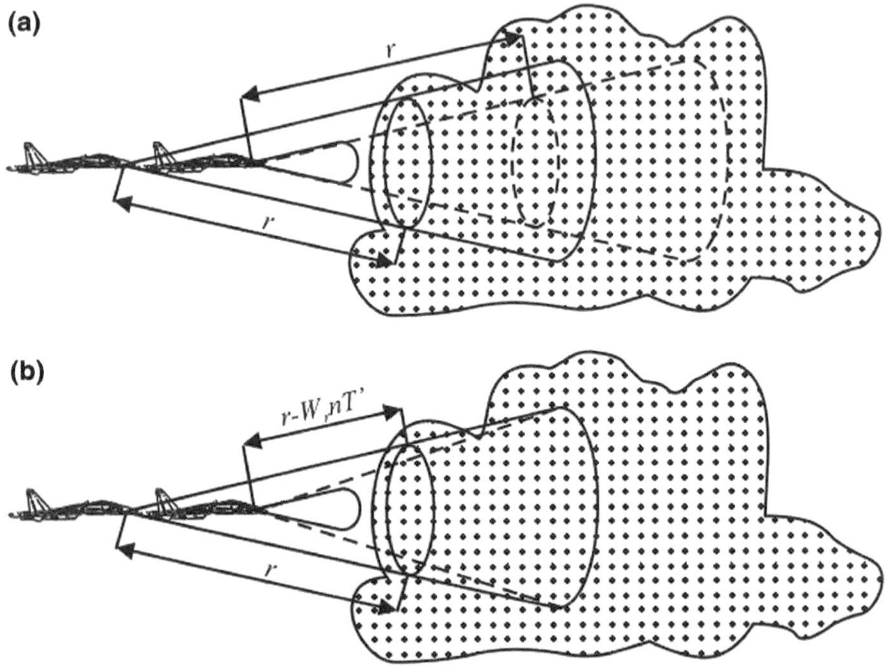

Fig. 1.3 Effect of carrier movement in the radar Doppler spectrum type reflected signals MO

$\vec{W}$	the vector of travelling speed of the RS carrier;
$W = \sqrt{W_r^2 + W_\tau^2}$	the module of travelling speed of the RS carrier;
θ_0	the angle between an axis of the pattern and the vector of the speed of the RS carrier;
$W_\tau = W \sin\theta_0$	the tangential component of the speed of the RS carrier;

$$\cos\theta_0 = \cos\alpha_0 \cos\beta_0; \quad \sin\theta_0 = \sqrt{\sin^2\alpha_0 + \cos^2\alpha_0 \sin^2\beta_0}.$$

Besides, the high radial velocity of the movement of the RS carrier (at small MO elevation angles) also leads to a considerable change of structure of reflectors in the allowed MO volume during the processing of signals [23]. It is caused by the fact that at the calculation of spectrum parameters, the readings of the reflected signals evenly remote from each other through the period of repetition T_p, that is defined in the timepoints fixed concerning the PS radiation moment are used. At the same time, signals from the allowed volume which is at the fixed distance from an RS, that is moving together with an RS are processed (Fig. 1.4a).

The change of spatial position of the allowed volume at the movement of the RS carrier leads to the change of structure of the reflectors forming a signal from V_i, and,

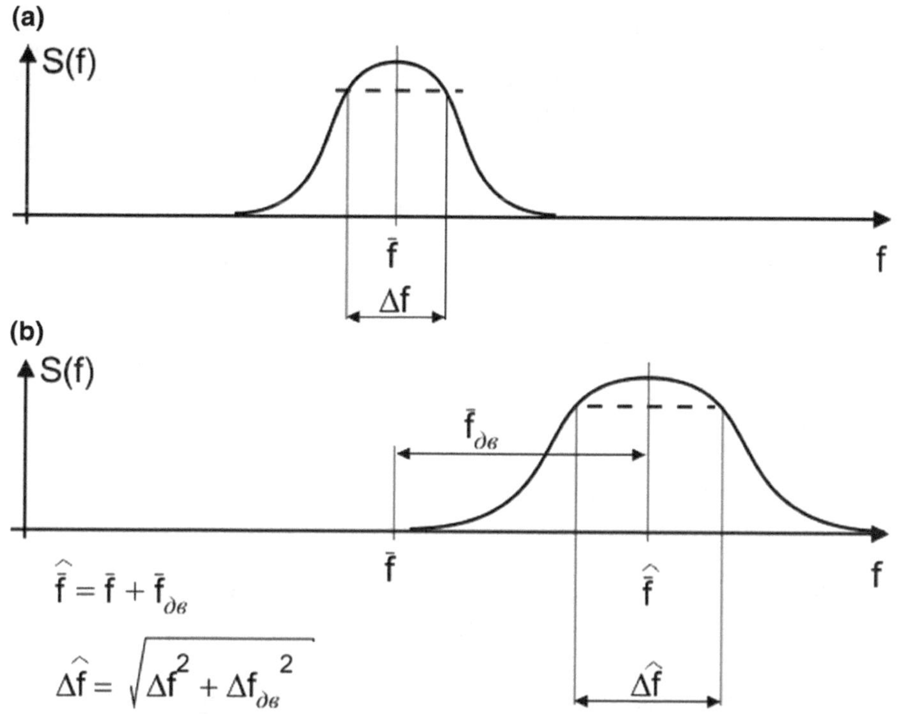

Fig. 1.4 Change composition resolvable radar volume upon movement of the support: and **a** fixed range; **b** when the moving distance

therefore, to the decorrelation of the reflected signals which is expressed in widening of a spectrum of fluctuations of signals (so-called "fluctuation at the fixed range").

There is also another reason for the widening of DS of the signals accepted by a mobile RS. The regular cross (in relation to the radiation direction) movement of reflectors caused, for example, by the tangential movement of the RS carrier leads to it [39, 40].

For the exclusion of fluctuations of the signals arising because of radial and tangential movement of the RS carrier, it is necessary to stabilize the position of the allowed volume in space in relation to the RS APC (Fig. 1.4b) or, in other words, to displace APC so that its position in relation to the centre of the allowed MO volume during the time of processing of the reflected signals would remain constant (the so-called RS "quasimotionless" mode).

Summing up the result, it should be noted that while using for detection within cloudy MO zones of heavy WS and intensive atmospheric turbulence, and while the assessment of their danger of an RS of 3-centimetre range of EMW which is considered optimum [41, 42] for the solution of the specified task, we receive the following results:

1. MO represents a spatially (volumetrically) distributed R target consisting of a set of a large number of accidentally located and independently moving elementary reflectors—HM. The information on their movement contains in DS of the reflected signals; therefore, for the assessment of the degree of danger, an RS should define the three first moments of a spectrum of signals in each allowed volume.
2. Own movement of the RS carrier leads to additional movement of MO reflectors along RS APC and, therefore, to significant distortion of the valid spectrum of their radial velocities. Therefore, during the detection of dangerous MO zones, it is necessary to compensate for the own movement of the RS carrier.
3. The task of the MO detection and the assessment of degree of its danger actually come down to the task of detection of the R signal reflected by MO against the background of noise and also the assessment of parameters of its DS, and for the reliable solution of this task, it is necessary within the allowed volume to provide the SNR not less than 28–30 dB.

1.3 The Requirements to Airborne RS at the Detection of Zones of Dangerous Wind Shears and Turbulence

It is possible to include [40] the view zone, the view period, the resolution on the range, angular coordinates and velocity, the accuracy of determination of coordinates and parameters of movement into the RS tactical characteristics significantly influencing the efficiency of dangerous MO zone detection.

The RS type and parameters, the DP form, RS energy potential and dynamic range of the receiver belong [40] to the RS technical characteristics on which the efficiency of the detection of dangerous MO zones depends.

1.3.1 The Requirements to a Zone, Method and Period of RS Review

The RS view zone intended for detection of the MO dangerous areas should provide a possibility of their preliminary detection and bypass. The borders of the view zone are defined by the RS range of radar station and view sectors on azimuth and elevation angle.

The RS range (the range of the detection of potentially dangerous MO) taking into account the values of AV flight velocity, admissible overloads, time of detection and intervention of crew in AV control, time of delay of draught entrance to the nominal mode should be about 200 km [3, 43, 44]. For supersonic AV, this value increases up to 600 km. On the other hand, the detection and assessment of WS danger are often made in the presence of MO with low specific EER and also in the conditions of intensive reflections from the underlying surface (US) on side lobes of DP. Therefore an RS should find areas of dangerous WS at the range of 10–20 km [27].

The RS antenna system (AS) on the majority of modern AV is placed in an AV fuselage nose cone. Therefore, the implementation of the sector view in a front hemisphere (the sector on azimuth ±(90–100)°, on elevation angle ±25°) is practically possible [43]. At a bigger deviation of DP on an elevation angle, the power of the signal reflected by US increases significantly.

The view velocity on angular coordinates is defined by necessary quantity of impulses M which should be received from the target for one pass of DP for provision of the required probabilistic characteristics of detection of the target and measurement of its parameters. The value of M depends on the period of repetition of T_n PS and the time of observation (analysis) of the target T_A.

The period of the single view of the whole zone is equal [30] to

$$T_{обз} = T_A N_A,$$

where $T_A = M T_n / N_r$;

$N_A = N_\alpha N_\beta N_r$	the quantity of resolution elements (the allowed volumes);
$N_r = r_{\max}/\delta r$	the quantity of resolution elements on the range;
$N_\alpha = \Phi_\alpha/\delta\alpha$	the quantity of resolution elements on azimuth;
$N_\beta = \Phi_\beta/\delta\beta$	the quantity of resolution elements on elevation angle;
Φ_α, Φ_β	the sizes of the view sector on azimuth and elevation angle, respectively.

On the other hand, the view period at the sector view is defined by the size of the sector of scanning and angular velocity of scanning [40]

$$T_{обз} = k_{обз} \frac{\Phi_\alpha}{\Omega_\alpha} \frac{\Phi_\beta}{\delta\beta},$$

where

$k_{обз}$	the coefficient considering losses of time for a beam reverse (at mechanical scanning $k_{обз} \approx 1{,}9$, and at electronic scanning $k_{обз} = 1$);
Ω_α	the angular speed of scanning in azimuth.

At $\delta\alpha = \delta\beta = 2,6^\circ$ [45], $\Phi_\alpha = 180^\circ$, $\Phi_\beta = 50^\circ$ [27], $r_{\max}/\delta r = 512$ [26], the total number of resolution elements $N_A = 800\,000$.

In case, if $M = 32$ and $T_n = 1$ ms then $T_{обз} = 50$ s.

1.3.2 The Requirements to the RS Resolution

The RS resolution on the range and angular coordinates limit the area of space within which the targets are not allowed; such area of space is called the allowed volume V_u.

The resolution on the range is determined by the spatial radius of correlation of fields of MO radial velocity and the radar reflectivity (RR) [46]. The spatial radius of correlation of MO radial velocity is defined by the external scale of zones of turbulence and WS which should be found by means of an RS. If the RS resolution on range exceeds the specified size, then reduction of the observed intensity of turbulence (smoothing of DS) takes place. RR is characterized by much faster spatial decorrelation than radial velocity. The average values of radii of correlation of velocity and RR are about 2 km and 300 m, respectively [46]. The appropriate lifetimes of not uniformity of spatial fields are equal to 500 s for velocities and less than 200 s for RR. Thus, the resolution of an airborne RS in the mode of detection of dangerous MO zones should be about 100 m [47].

The RS resolution on angular coordinates depends on the coherence of RS PS and a type of interperiod processing of the reflected signals. At incoherent processing, the RS resolution on angular coordinates is determined by the DP width on power at the level of −3 dB in the plane of reafing of the corresponding coordinate [43, 48], and at coherent processing it can be estimated by the duration of a response of the coordinated filter [49, 50]

$$\delta\alpha = -|\mathrm{tg}\,\alpha| + \sqrt{\mathrm{tg}^2\alpha + \frac{\lambda}{W T_H \cos\alpha \cos\beta}},$$

where

W the ground speed of the RS carrier;
T_N the time of coherent accumulation.

For the spatial localization of dangerous for flights zones of strong wind shears and intensive turbulence the angular RS resolution should be about 1–3° [51].

1.3.3 The Requirements to the RS Accuracy

The RS accuracy characteristics depend on the mode of its functioning. In the mode of detection of the signal reflected by MO, the accuracy of detection is defined by the probability of false alarm p_F and the probability of the admission of a signal p_D. In the mode of detection and assessment of the degree of danger of the WS areas and atmospheric turbulence on the basis of the analysis of DS of the reflected signals, the following accuracy of measurement should be provided:

- of range and angular coordinates—at the level of the RS resolution [43], i.e. about 100 m on range and 1–3° on angular coordinates;
- of radial speed—not more than ±1 m/s as quality indicators of dangerous MO are defined with such discretization (Table 1.1).

1.3.4 The Requirements to Parameters of the RS Probing Signal

For detection and measurement of the MO parameters, the pulse method of radiation (modulation) is used as a rule [43]. Its advantages are division on time of the moments of PS radiation and reception of the reflected signals, the possibility of exclusion of influence of the transmitter on an RS reception path, convenience of synchronization of paths of reception and transfer, convenience of separation and fixing of information on range and also high resolution on the range. The disadvantage of pulse PS—its small power consumption in comparison with a continuous signal—can be compensated by radiation frequency with the subsequent accumulation of the reflected signals.

The PS parameters defining the RS characteristics at detection and assessment of the degree of MO danger include duration, period (frequency) of repetition and coherence degree.

Duration of the impulse PS τ_u with rectangular envelope at a bell-shaped form of the amplitude–frequency characteristic (AFC) of filters of the intermediate frequency amplifier (IFA) of the RS receiver is inversely proportional to the width of a modulation spectrum of PS Δf_M [48: $\tau_u \approx 1/\Delta f_M$.

The resolution of an RS on the range is inversely proportional to the value of Δf_M and is defined by the expression

$$\delta r = c/(2\Delta f_M).$$

The maximum allowed duration of PS is connected with the resolution on the range by the expression

$$\tau_u = 2\delta r/c.$$

At the RS-fixed pulse power, the increase in PS duration with the corresponding proportional narrowing of the bandwidth of the RS receiver IFA provides an increase in range of detection of potentially dangerous MO. However, at the same time, firstly, the resolution and accuracy of assessment of range worsen significantly, and secondly, possibilities of RS on the detection of dangerous zones of small-scale atmospheric turbulence and WS decrease. Characteristic values of RS PS duration for the mode of detection and assessment of the degree of danger of zones of turbulence and WS are values in the range of 0.33–1.0 ms [51, 52].

The coherence of R signals can be provided in various ways, however, in most perspective state and civil airborne RS, the truly coherent scheme of RS construction is implemented. At the same time, the coherent oscillator (CO) is a source of a basic signal. The requirements to short-term frequency stability of CO are defined by the necessary accuracy δV of measurement of the radial speed of the targets

$$\Delta f_{z}/f_{z} \leq \sqrt{2}\delta V/c,$$

where

f_{z} the frequency of the signal developed by CO;
Δf_{z} the change of the value of f_{z} during repetition of RS PS.

At $\delta V = 1\ \mathrm{m/s}$, the relative change of CO frequency during repetition should not exceed the values of 10^{-9}–10^{-8} [53].

Long-term frequency drifts can be considerable and should be adjusted by the scheme of the automatic-frequency control (AFC).

Depending on the pulse rate of radiation, truly coherent RS can work in the mode of high (Q > 10) or small (Q > 10) pulse rate. In the latter case, they are also called pulse-Doppler (quasicontinuous). Depending on the used frequency of PS, repetition of pulse-Doppler RS can be with low (LFR), average (AFR) or high (HFR) frequency of repetition [54]. The value of frequency of repetition can be defined from the condition of unambiguous measurement of range to the target on the view zone border

$$f_n = c/2r_{\max}.$$

As the RS coverage in the meteo, turbulence and wind shear modes is different, the values of frequency of repetition of the RS probing signals will also differ. In particular, using LFR is preferable in the meteo mode for the exclusion of uncertainty on the range, and in the turbulence and wind shear modes for the purpose of reduction of uncertainty on the range at the preservation of a wide range of unambiguously determined velocities, it is necessary to apply AFR.

1.3.5 The Requirements to the Pattern of the RS AS

Characteristic of the airborne RS intended for detection and assessment of the degree of MO danger is the use of highly directional (needle) DP. Its width in the azimuthal and elevation planes on the AS corresponding linear sizes d_α, d_β [43]

$$\Delta\alpha = (60\ldots70^{\circ})\lambda/d_\alpha, \quad \Delta\beta = (60\ldots70^{\circ})\lambda/d_\beta.$$

It is impossible to significantly reduce the DP width, since this reduces the number of pulses reflected by the MO in the processed pack, which reduces the probability of detecting dangerous MO. Using the AS with the diameter of 762 mm [26, 27], $\Delta\alpha = \Delta\beta \approx 2,6^{\circ}$.

The important parameter of an RS is the maximum level of side lobes (LSL) of a DP [55]

$$\eta = \max\{10\lg[G(\alpha_i, \beta_i)/G(\alpha_0, \beta_0)]\},$$

where

α_0, β_0 the angular coordinates of the DP maximum;
α_i, β_i the angular coordinates of the DP ith side lobe.

LSL at detection of dangerous MO should not exceed −23 dB [51].

1.3.6 The Requirements to the RS Energy Potential

The main measure of RS power opportunities is its energy (meteorological) potential [26, 27, 43]

$$\varPi_{M} = P_u K_g^2 \tau_u / P_{\min}, \quad (1.24)$$

where

P_u the RS PS pulse power;
K_g the AS amplification factor;
$P_{\min}$ sensitivity of the RS receiver.

The expression (1.24) can be written down in a logarithmic form

$$\varPi_{M}[\partial Б] = 10\lg P_u + 20\lg K_g + 10\lg \tau_u - 10\lg P_{\min}. \quad (1.25)$$

The minimum permissible values of energy potential (1.25) depending on the RS maximum coverage of and velocity of its carrier are given in Table 1.2 [27]. The provided data correspond to the 3-centimetre range of RS operating wavelengths.

1.3.7 The Requirements to the Dynamic Range of the RS Receiver

Unlike the modes of detection of targets of RS, in the mode of the assessment of the degree of MO danger not minimum found signal, and the accuracy of measurement

Table 1.2 Minimum permissible values of the RS energy [27]

Cruising speed of an AV, km/h	RS coverage, km	$\Pi_{M\ \min}$, dB
Less then 200	50	167
200–400	100	179
400–650	150	187
650–925	200	192
925–1200	250	197
More than 1200	300	201

of the DS parameters of the reflected R signals in all dynamic ranges which can reach 110–30 dB, [56] is important. It results in great difficulties at the construction of an RS receiving path which usually has two channels: the logarithmic channel with a big dynamic range for RR measurement and the linear channel with a limited dynamic range for measurement of the DS parameters of the reflected signals. At the same time at an input of the linear channel of the RS receiver, the fast automatic level control (ALC) operated on a signal from the output of the logarithmic channel and allowing to adjust automatically, the working site of an amplitude characteristic of the linear channel to the average level of the reflected signal in a concrete resolution element is usually applied. Besides, for exact measurement of RR value, it is necessary to normalize constantly the power of the received signal in all range bracket.

1.4 Condition of Developments of the Domestic and Foreign RS Providing the Assessment of Zones of Dangerous MO

1.4.1 The Analysis of the Condition of Researches on Determination of the MO Parameters by Land RS

A large number of researches on the use of RS of microwave spectrum for the assessment of the various MO parameters is carried out so far. The main achievements in this area belong to scientists of the Russian Federation (earlier the USSR) and the United States.

The fundamental works in this field are the works of the employees of the Main Geophysical Observatory (MGO) named after Voyeykov (St. Petersburg) and St. Petersburg SMI Stepanenko, Melnik, Brylev, Melnikov, and Central Aerological Observatory (CAO) (Dolgoprudny, Moscow Region) Chernikov, Gorelik, Melnichuk, etc. [7, 53, 57–60]. In the specified works, the possibility of use of RS of different modifications for the assessment of various MO parameters and, in particular, the connection of characteristics of the reflected R signal with the MO physical parameters, such as distribution of drops by the sizes, intensity of rainfall, aggregate state of MO particles, movement of particles in clouds and rainfall is analyzed. The main attention at the same time was paid to amplitude characteristics of the reflected signals which are connected with the MO RR.

The following works should be noted from the foreign works in the field of processing of signals of meteorological RS: the works of Atlas [61], Battan [62], Bing and Dutton [63], Smith et al. [56] and Doviak and Zrnich [46, 64].

In the specified works at the analysis of power characteristics of the R signals reflected by MO, the decision on the degree of danger of this or that site of MO was made on the basis of a brightness picture of RR spatial distribution. At the same time, on the basis of the processing of results of a large number of meteorological radar observations, a number of empirical criteria for the assessment of the degree of MO danger was developed. One of such criteria is the RR horizontal gradient *dZ/Dr*. It was offered to use the parameter

$$Y = h\frac{\lg Z_{\max}}{\lg Z_{\min}} - \lg Z_i,$$

where

h the MO height;
Z_i RR at the level exceeding the area of maximum reflectivity by 2 km.

However, it should be noted here that RR characterizes only the degree of MO volume saturation by water and is an indirect sign of the existence of dangerous zones, especially zones of turbulence [53, 65].

For the expansion of a circle of solvable problems, the transition to using information contained in spectral characteristics of the signals reflected by MO was natural. Up to the mid-1960s, the Soviet scientists Melnichuk et al. [53] and the American researcher Lermitt [66] proved the possibility of use of pulse-Doppler RS at the research of the processes happening in MO volume that allowed to investigate dynamics of these processes. In the work of [66] R. Lermitt for the first time defined the requirements to pulseDoppler RS for the analysis of MO characteristics. The incoherent methods of Doppler (interperiod) processing of signals are the subject of numerous works of employees of MSHO [67, 68] and CAO [53, 69, 70] and also the American researcher Atlas [17, 61].

From the beginning of the 1970s, the intensive use of coherent RS for the analysis of dynamics of MO development began. This issue was the subject of numerous works of the American scientists Sirmans et al. [46, 64, 66], in which the problems of the organization of information processing in real time by both single-pulse-Doppler RS and systems of several RS are covered and also the works of the Soviet scientists [68, 71, 72] in which the analysis of pulse-Doppler RS up to the modes with the synthesized antenna aperture was made. In the specified works, the significant expansion of opportunities of meteorological RS when using coherent processing of signals is emphasized. However, the vast majority of the specified researches belongs to the land RS of all-round coverage intended for the purposes of the meteonotification service. At the same time, all potential opportunities put in coherent radio signals were often not used, and processing came down to the analysis of properties of a signal envelope [67, 73].

Besides, in the domestic literature, now there are practically no publications devoted to the issues of creation of the Airborne Doppler RS with digital information processing in real time providing an opportunity, along with traditional methods of assessment of RR, to carry out the analysis of dynamic processes in moisture target volume, in particular, to estimate a degree of danger of such phenomena as turbulence and the wind phenomena.

1.4.2 The Analysis of the Condition of Domestic and Foreign Developments of Airborne RS Regarding Detection and Danger Assessment of MO

The main directions of foreign developments in the field of creation of airborne RS for civil AV are set by the available regulatory requirements of the international industry organizations. In particular, the main requirements for the airborne radio-electronic equipment (AREE) are regulated by the documents of *ICAO*. At the same time, the provisions of the normals of *ARINC* Corporation (USA) are used as the basic ones.

The normal ARINC-708 (from 04.05.1978) [26] provides the realization in airborne meteo RS of the modes of the definition of dangerous situations like "meteo" and "turbulence". The requirements to the key technical parameters of airborne RS formulated in the normal ARINC-708 are provided in Table 1.3. The main differences

Table 1.3 The key technical parameters of an RS of an AV of civil appointment [26]

Parameter	Value
Frequency range, MHz	9345±20; 9375±20; 5400
Coverage, km	0–590
Field of view, deg.: – in the horizontal plane – in the vertical plane	 ±90 ±14
Number of scannings per minute	15
Accuracy of stabilization of an antenna, deg.	±0.5
The measured parameters	Azimuth, coverage, rain intensity
Resolution on range (the number of resolution elements at the maximum range)	128, 256 or 512
Accuracy of indication of: – angular coordinates, deg. – range, % of the measured distance	 ±2 ±4
Antenna system type	Flat-slotted-guide antenna array with a diameter of 762 mm
Level of side lobe, dB, not more than	−21 dB

of this normal from previous ones are the existence of the requirements:

- the sharp increase in stability of the setting generator, the transmitting device and an oscillator of the receiver of an RS for the purpose of formation of the coherent fluctuations necessary for Doppler processing of the received signals;
- provision of completely digital exchange of information between units of an RS and with external systems;
- the use of the microprocessor equipment;
- the increase in reliability and a decrease in weight of the equipment.

The normal *ARINC-708A* (from 27.12.1993) [27] in addition introduced the mode of WS detection (Table 1.4).

The specified normals of ARINC corporation defined technical characteristics of civil AV equipment. The requirements of operational characteristics of airborne meteo RS are defined by the *RTCADO-220* [74], *RTCA DO-173* [75] and *RTCA DO-178B* [76] standards developed by the International Radio Technical Commission for Aeronautics (RTCA).

The provided normative documents unambiguously indicate the necessity of realization of the meteo, turbulence and wind shear modes in an airborne meteo RS of a civil AV, passing the international certification. In this regard, most manufacturing firms of avionics for civil AV offer the airborne meteo RS representing the coherent pulse-Doppler systems with the solid-state transmitter and the AS in the form of the flat waveguide slot array antenna (WSAA) with horizontal polarization conforming to the requirements of normals of ARINC. The American firms *Rockwell Collins* and *Honeywell Aerospace Inc* dominate now in this sector.

The modern coherent *RLSWXR-2100 MultiScanThreatTrack*™ *Version 2.0* of *Rockwell/Collins* (USA) is intended for installation on the long-haul AV [44]. The RS developed according to the requirements of normals *ARINC-708* and *ARINS-708A,* automatically analyzes an air situation on a route, predicts existence of zones

Table 1.4 Technical requirements to RS of civil AV for the realization of the mode of wind shear detection [27]

Requirement	Value
The sector of wind shear detection, not less than, deg.	±25
The resolution on the range at detection of dangerous wind shear, not worse than, km	0.37
The range of development of a signal "Warning about a dangerous wind shear" (Level 3 Alert), km	0.46–2.78
The range of development of a signal "Warning about a dangerous wind shear" (Level 2 Alert), km	2.78–5.56
The range of development of a signal "Information about a dangerous wind shear" (Level 1 Alert), km	5.56–9.26
The dynamic range of the RS receiver of the radar station, not less than, dB	60
Accuracy of stabilization of an antenna, deg.	±0.5
Level of side lobe, dB, not more than	−25 dB

of dangerous MO (zones of the increased turbulence, wind shear, etc.) and forms on the display the map of dangerous MO without the need for manual control of operating modes. The integrated system built using modern microelectronic microwave technologies, contains the solid-state transmitter with the pulse power of 150 W (Table 1.5), the receiver with quartz stabilization of frequency of an oscillator and a low-noise input cascade on field transistors made of gallium arsenide. For the provision of the required coverage, the increase in the duration of the probing impulse, increase in receiver sensitivity, reduction of its bandwidth (up to 70 kHz)

Table 1.5 The main functional capabilities of the civil airborne RS produced by the leading foreign producers

Name	WXR-2100	RDR-4000
Object of installation	Long-haul AV of the Airbus A380, Boeing-767, Boeing-777 type and similar to them	
Developer	Rockwell/Collins	Honeywell Aerospace
Coverage, km		
– in the Meteo mode	600	600
– in the Turbulence mode	75	110
– in the Wind shear mode	10	10
Field of view, deg.:		
– on azimuth	±90	±90 (in the WS mode ±40)
– on an elevation angle	±40	–
Transmitter		
Power, W:		
– average	–	–
– impulse	150	–
Duration of an impulse, microsec	1–25	–
Frequency of repetition, Hz	180 (up to 3000)	–
Receiver		
Noise ratio, dB	3.8	1.9
Frequency band, MHz	32	
Sensitivity, dBm	−125	−124
Antenna system		
Type	SGAA, Ø762 mm	SGAA, Ø 305–762 mm (4 variants)
Augmentation ratio, dB	34.5	28,5…34,8
Pattern width, deg.		
– on azimuth	3.5	3,0…8,0
– on an elevation angle	2.5	–
LSL, dB	−31	–
Scanning velocity, deg./s	45	90

and increase in dynamic range are provided in the system. The manufacturing firm offers *RLSWXR-2100* for modernization of an AV equipped with *WXR-700* systems. Besides, the *Rockwell/Collins* company offers a number of the civil and state systems providing a mapping of US along a flight route, detection of MO with big specific EER and also detection of dangerous zones of turbulence (*TWR-850/WXR-840/WXR-800, RTA-4200*, etc.).

Honeywell Aerospace Inc. offers airborne RS of new generation *RDR-4000 IntuVue*™ *3D* widely using the technical solutions which were earlier applied only in the state systems (PS compression, suppression of reflections from US, storage in the memory device of the results of R observation of the whole front hemisphere, etc.). RS differ significantly reduced weight and size (by 50–60%) of the processing unit, the improved AC drive (providing greater accuracy and speed of review), modified software (providing mapping of the surface, detection of dangerous WS and turbulence with greater accuracy and reliability). The functional capabilities of this RS are also given in Table 1.5.

It should be noted that the leading foreign firms (*Rockwell Collins, Honeywell Aerospace Inc.*, etc.) together with the public and non-profit Russian institutions (*NASA, ARINCI*, etc.) do not stop at the reached level. Now they carry out works on the improvement of characteristics, creation of new hardware, the algorithmic and special software for the airborne meteo RS, significantly improving their characteristics during the operation in the modes of detection of strong wind shears and intensive turbulence [8, 9, 77].

In domestic practice, the requirements to the airborne radio-electronic equipment of a civil AV are defined by the annex "Technical Requirements to the Plane Equipment" and "Uniform Standards of the Flight Worthiness of Civil Transport Planes" [78]. The Sect. 2 8.3.9 of the specified requirements is made by "Technical Requirements to Meteonavigation Radars". In particular, the obligatory modes of airborne meteo RS for AV of domestic production are the meteo and turbulence modes. The wind shear mode is not obligatory. However, the representatives of state bodies responsible for the organization and safety of flights consider now that the presence of such mode would contribute significantly to AV FOS.

Most the existing domestic meteonavigation RS are built according to the incoherent scheme because of the existence of significant technical and economic restrictions [44]. The incoherent scheme does not allow using ful information on an air situation is contained in a phase structure of the reflected signals. At the same time, the decision on the danger of this or that MO is made by an AV crew on the basis of the analysis of a brightness picture of the spatial distribution of MO RR, i.e. actually on the basis of the analysis of average power of the reflected signal [79].

However, at the Russian market, there are also modern hi-tech systems meeting the requirements of the normal ARINC-708A. For example, JSC Kotlin-Novator (St. Petersburg) produces the airborne coherent meteonavigation RS A882 providing the ranges of detection of the following:

- large storm fronts and powerful thunderstorms—400–600 km;
- thunderstorms and storm activity—300–400 km;

- cumulus and rain activity—250–300 km;
- zones of the increased turbulence—90 km;
- areas of dangerous wind shear—10 km;
- the large cities (Moscow), coastal line—360 km;
- peaks of mountain tops—up to 250 km;
- planes (Il-76 type)—30–40 km.

The Contour-10SV system possessing similar functional capabilities is produced by LLC Contour-NIIRS (St. Petersburg). The RS carries out the following:

(a) detection of convective MO (thunderstorms, powerful cumulus cloud cover) with the definition of degree of their danger and also the zones of dangerous turbulence in MO;
(b) display of the MO vertical profile on the chosen direction and also supports the mode of three-dimensional indication of a meteosituation;
(c) detection of WS zones in which the F-factor exceeds a certain value, according to the DO-220 requirements;
(d) detection of characteristic land reference points (a city, industrial constructions, large reservoirs and coastal line, vessels on a water surface, etc.).

Unlike meteo and meteonavigation RS installed on civil AV for which the problem of detection of dangerous zones MO is one of the main, the airborne multimode RS for transport aircraft used primarily as means of information support of the solution of various navigation tasks [49, 80, 81]. However, at the solution of any of them the problem of ensuring the flight of an AV round the clock in any weather conditions is important for what the AV crew should receive in real time objective and reliable information about location of MO zones dangerous for flights; therefore now the issue of use in airborne RS of algorithms of processing of the signals reflected by MO which would allow to carry out the effective detection of the dynamic phenomena, the most dangerous for flight (intensive turbulence, WS, etc.) came up.

In particular, in the United States, the airborne RS *AN/APG-77* with the active phased antenna array (APAA) with the diameter about 1 m containing about 1500 active transmit–receive modules is designed to become the main airborne avionics of perspective AV of the 5th generation *F-22*. As one of the main operating modes of *AN/APG-77* RS [82] along with the modes providing target use of arms, the developers of this perspective system allocated the mode of the definition of weather conditions along a flight route (*Weather*/Meteo mode).

The domestic airborne RS operated now as a rule have no modes "turbulence" and "wind shear". It is expedient to realize the specified modes as a part of multimode airborne RS of the front view at the minimum completion of its hardware. At the same time the development of the modes comes down to the choice of the PS parameters and AS characteristics and also to the creation of algorithmic and software of computing means including algorithms of processing of R information and management of work of the station.

The RS of the front view with FAA providing formation and scanning of a "sharp beam", "wide beam", "flat beam", "cosecant beam" and "focused beam" DP allows to find dangerous areas of atmospheric turbulence.

1.5 Improvement of Methods of Processing of MO Signals in Airborne RS

Now the methods of processing of the signals reflected by MO in airborne RS are being intensively developed both on the way of using more perfect technical means and on the way of development of new improvement of the existing methods of processing and interpretation of results of R observations [35]. At the same time, there is a number of problems in the determination of the RS structure considering specifics of its airborne use in the choice of the algorithms of processing of the signals, best considering the advantages of coherent and incoherent methods of reception and also giving the chance of operational assessment of degree of MO danger taking into account its fast variability. In particular, the feature of using regular RS of the state AV focused on the solution of specific tasks, unlike airborne RS of planes of civil aviation; it is a fact that MO is assumed by sources of the interfering signals complicating the process of detection, classification and accompaniment of air targets. Therefore, the principles of construction of RS of air fight, the modes of their use and information processing are directed, first of all, to selection and suppression of the specified signals. For decrease in the influence of MO on an RS operation, a number of measures of increase in noise immunity is provided: temporal ALC, the moving-target selection (MTS), the selection on spatial coordinates, polarizing selection and others. At the same time, in solving the problems of detection and assessment of the danger of the areas of WS and atmospheric turbulence, the interference that must be suppressed will be signals from high-speed pinpoint targets. Besides, the own movement of the RS carrier leads to additional movement of all R targets which are in a view zone in relation to RS APC and, therefore, to significant distortion of the valid spectrum of their radial velocities. Therefore, at the assessment of a degree of danger of MO zones, it is necessary to compensate for the movement of an RS carrier during processing.

Thus, the purpose of researches is the development of methods and parametric algorithms of digital coherent processing of signals of the airborne RS located on an AV on the basis of the autoregression (AR) model, providing increase in accuracy of danger assessment for flight of WS and areas of atmospheric turbulence and also definition of ways of their implementation.

For the achievement of the aim, it is necessary to solve the following main objectives:

- to develop mathematical models of the signals reflected from MO in the conditions of WS and turbulence and received by the RS installed at the AV board;
- to synthesize the algorithms of digital coherent processing of signals for detection and danger assessment of the areas of WS and turbulence;
- to estimate the characteristics of the developed algorithms by mathematical modelling;
- to estimate the influence of the destabilizing factors arising at RS carrier flying on the efficiency of processing of signals and to develop algorithms of compensation of this influence;

– to define the ways of implementation of the developed algorithms on the basis of the modern hardware and software recommended for use in airborne RS. To develop recommendations about the specification of the requirements to airborne RS of AV regarding the modes of detection of dangerous MO zones.

The Main Conclusions for Sect. 1.1

1. Detection of MO with the assessment of danger degree of strong WS and intensive turbulence zones is one of the main operating modes of an airborne RS assuming the automatic detection and measurement of the DS parameters of the signals reflected by MO.
2. Own movement of the RS carrier leads to additional movement of MO reflectors along RS APC and, therefore, to significant distortion of the valid spectrum of their radial velocities; therefore at detection of dangerous MO zones, it is necessary to compensate the own movement of the RS carrier.
3. The need of accurate, reliable and timely assessment of components of wind speed on the MO volume imposes strict requirements to the characteristics of MO signal detection (the probability of the correct detection $p_D \geq 0{,}9$ at the probability of false alarm $p_F \leq 10^{-4}$ and the accuracy of measurement of the DS parameters of an MO signal (an RMS error in terms of velocity is not more than 1 m/s) whose achievement is possible at provision of the SNR not less than 28–30 dB.
4. The purpose of researches is development of methods and parametric algorithms of digital coherent processing of the radar signals on the basis of autoregression model providing increase in accuracy of the assessment of danger of areas of WS and atmospheric turbulence for an AV flight and also definition of ways of their implementation on the basis of the hardware and software of perspective airborne RS.
5. One of the main methods of researches is mathematical modelling. The choice of this method is caused by the absence of prior information on parameters and statistical properties of a useful signal and interferences.

References

1. Astapenko PD, Baranov AM, Shvarev IM (1980) Weather and flights of planes and helicopters. L. Hydrometeoizdat, 280 pages
2. Flight operation safety / Eds. R.V. Sakach. - M.: Transport, 1989. - 239 pages
3. Filatov G.A., Puminova G.S., Silvestrov P.V. Flight operation safety in the turbulent atmosphere. - M.: Transport, 1992. - 272 pages
4. The guide to forecasting of weather conditions for aircraft / Eds. K.G. Abramovich and A.A. Vasilyev. - L.: Hydrometeoizdat, 1985. - 301 pages

5. Uskov V. Wind shear and its influence on landing. Civil aviation, 1987, No. 12, pages 27-29
6. Research of heterogeneity of the wind field in clouds and rainfall by means of the automated incoherent radar / Ivanova G.V., Malanichev S.A., Melnik Yu.A., Melnikov V.M., Mikhaylova E.I., Ryzhkov A.V. Works of GGO, 1988, issue 526. P. 16–22
7. Melnik Yu. A., Melnikov V.M., Ryzhkov A.V. Possibilities of use of the single doppler radar in the meteorological purposes: Review, Works of GGO, 1991, issue 538. P. 8–18
8. Bowles RL (1990) Windshear detection and avoidance—airborne systems survey. In: Proceedings of 29th IEEE conference on decision and control, vol. 2, pp 708–736
9. Proctor FH, Hinton DA, Bowles RL (2000) A windshear hazard index. In: Preprints of 9th conference on aviation, range and aerospace meteorology (11–15 Sept 2000), paper 7.7, pp 482–487
10. Aviation meteorology: Textbook / A.M. Baranov, O.G. Bogatkin, V.F. Goverdovsky, et al..; Eds. A.A. Vasilyev. - SPb.: Hydrometeoizdat, 1992. - 347 pages
11. Matveev L.T. Course of the general meteorology. Physics of the atmosphere. - L.: Hydrometeoizdat, 1984. - 751 pages
12. Sviridov N. In the turbulent atmosphere. Civil aviation, 1978, No. 9, pages 44–46
13. Turbulence in the free atmosphere / N.K. Vinichenko, N.Z. Pinus, S.M. Shmeter, G.N. Schur. - L.: Hydrometeoizdat, 1976. - 288 pages
14. Introduction to aero autoelasticity. / S.M. Belotserkovsky et al. - M.: Science, 1980 – 384 pages
15. Vasilyev A.A., Leshkevich T.V. Atmospheric turbulence and bumpiness of aircrafts. - In the book: Weather conditions of flights of aircrafts at small heights. - L.: Hydrometeoizdat, 1983. - p. 51–61
16. Ostrovityanov R.V., Basalov F.A. Statistical theory of a radar-location of the extended targets. - M.: Radio and Communication, 1982. - 232 pages
17. Atlas D., Srivastava R.C. A Method for Radar Turbulence Detection. IEEE Transactions on Aerospace and Electronic Systems, 1971, v. AES-7, [1]1, p.p. 179–187
18. Van Tris G. Theory of detection, estimates and modulation. The translation from English - M.: "Sov. Radio", 1977 - v. 3. - 664 pages
19. Shirman Ya.D., Manzhos V.N. The theory and technology of processing of radar information against the background of interferences. - M.: Radio and Communication, 1981. - 416 pages
20. Shlyakhin V.M. Probabilistic models of not Rayleigh fluctuations of radar signals: Review, Radio engineering and electronics, 1987, v. 32, No. 9, pages 1793-1817.
21. Krasyuk N.P., Rosenberg V.I. Ship radar-location and meteorology. - L.: Shipbuilding, 1970. - 325 pages
22. Feldman Yu.I., Gidaspov Yu.B., Gomzin V.N. Accompaniment of moving targets. - M.: Sov. Radio, 1978. - 288 pages
23. Feldman Yu.I., Mandurovsky I.A. The theory of fluctuations of the locational signals reflected by the distributed targets. - M.: Radio and Communication, 1988. - 272 pages
24. Sosulin Yu.G. Theoretical bases of radar-location and radio navigation: Ed. book. - M.: Radio and Communication, 1992. - 304 pages
25. Federal aviation rules of flights in airspace of the Russian Federation: Annex to the order of the Minister of Defence of the Russian Federation, Ministry of Transport of the Russian Federation and Russian Aviation and Space Agency from 31.03.02 No. 136/42/51
26. ARINC-708. Airborne Weather Radar. - Aeronautical Radio Inc., Annapolis, Maryland, USA, 1979
27. ARINC-708A. Airborne Weather Radar with Forward Looking Windshear Detection Capability. - Aeronautical Radio Inc., Annapolis, Maryland, USA, 1993
28. Adaptive spatial doppler processing of echo signals in RS of air traffic control / G.N. Gromov, Yu.V. Ivanov, T. G. Savelyev, E.A. Sinitsyn. - SPb.: FSUE VNIIRA, 2002. - 270 pages
29. Chernyshov E.E., Mikhaylutsa K.T., Vereshchagin A.V. Comparative analysis of radar methods of assessment of spectral characteristics of moisture targets: The report on the XVII All-Russian symposium "Radar research of environments" (20–22.04.1999). - In the book: Works of the XVI-XIX All-Russian symposiums "Radar research of environments". Issue 2. - SPb.: VIKU, 2002 - p. 228-239

30. Aviation radar-location: Reference book. / Eds. P. S. Davydov. - M.: Transport, 1984. - 223 pages
31. Zubkov B.V., Minayev E.R. Bases of safety of flights. - M.: Transport, 1987. - 143 pages
32. Adaptive radio engineering systems with antenna arrays. / A.K. Zhuravlev, V.A. Khlebnikov, A.P. Rodimov et al. - L.: LSU publishing house, 1991 - 544 pages
33. Tarasov I. Design of the configured processors on FPLD base. Components and technologies, 2006, No. 2, pages 78–83
34. Vereshchagin A.V., Ivanov Yu.V., Perelomov V.N., Myasnikov S. A., Sinitsyn V. A., Sinitsyn E.A. Processing of radar signals of airborne coherent and pulse radar stations of planes in difficult meteoconditions. St.-Petersburg, pub. house. Research Centre ART, 2016. - 239 pages
35. Melnikov V.M. Information processing in doppler MRL. Foreign radio electronics, 1993, No. 4, pages 35–42
36. Bakulev P.A., Stepin V.M. Methods and devices of selection of moving targets. - M.: Radio and Communication, 1986. - 288 pages
37. Trunk G.V. Coefficient of losses at accumulation of noise in the MTS systems. TIIER, 1977, t.65, No. 5, pages 115–116
38. Komarov V.M., Andreyeva T. M., Yanovitsky A.K. Airborne pulse-doppler radar stations. Foreign radio electronics, 1991, No. 9–10
39. Okhrimenko A.E. Bases of radar-location and radio-electronic fight: Text book for higher education institutions. P.1. Bases of radar-location. - M.: Voyenizdat, 1983. - 456 pages
40. Theoretical bases of a radar-location: Ed. book for higher education institutions /A.A. Korostelev, N.F. Klyuev, Yu.A. Melnik, et al.; Eds. V.E. Dulevich. - M.: Sov. Radio, 1978. - 608 pages
41. Ryzhkov A.V. About accounting of inertia of hydrometeors at measurement of turbulence by a radar method. Works of the 6th All-Union meeting on radar meteorology. – L.: Hydrometeoizdat, 1984.- Pages 132–136
42. Ryzhkov A.V. Meteorological objects and their radar characteristics. Foreign radio electronics, 1993, No. 4, pages 6–18
43. Radar systems of air vehicles: Textbook for higher education institutions / Eds. P. S. Davydov. - M.: Transport, 1977. - 352 pages
44. Radar systems of air vessels: The textbook for higher education institutions. / Eds. P. S. Davydov. - M.: Transport, 1988. - 359 pages
45. Sinani A.I., White Yu.I. Electronic scanning in control systems of arms of fighters, the World of avionics, 2002, No. 1, pages 23–28
46. Doviak R., Zrnich. Doppler radars and meteorological observations: The translation from English - L.: Hydrometeoizdat, 1988. - 511 pages
47. Melnikov V.M. Meteorological informational content of doppler radars. Works of the All-Russian symposium "Radar researches of environments". Issue 1. - SPb.: MSA named after A.F. Mozhaisky, 1997. - p. 165–172
48. Tuchkov N.T. The automated systems and radio-electronic facilities of air traffic control. - M.: Transport, 1994. - 368 pages
49. Guskov Yu.N., Zhiburtovich N.Yu. The principles of design of family of the unified multipurpose airborne radar stations of fighter aircrafts. Radiosystem, issue 65, p. 6–10
50. Jenkins G., Watts D. Spectral analysis and its applications: The translation from English in 2 v. - M.: Mir, 1971 –1972
51. Perevezentsev L.T., Ogarkov V.N. Radar systems of the airports: The textbook for higher education institutions. - M.: Transport. 1991. - 360 pages
52. RDR-4B. Forward Looking Windshear Detection / Weather Radar System. User's Manual with Radar Operating Guidelines. Rev. 4/00. - Honeywell International Inc., Redmond, Washington USA, 2000
53. Gorelik A.G., Melnichuk Yu.V., Chernikov A.A. Connection of statistical characteristics of a radar signal with dynamic processes and microstructure of a meteoobject. Works of the CAO, 1963, issue 48, pages 3–55

54. Mikhaylutsa K.T. Digital processing of signals in airborne radar-tracking systems: Text book / Eds. V.F. Volobuyev. - L.: LIAP, 1988. - 78 pages
55. Minkovich B.M., Yakovlev V. P. Theory of synthesis of antennas. - M.: Sov. radio, 1969. - 296 pages
56. Smith P., Hardy K., Glover K. Radars in meteorology, TIIER, t.62, No. 6, 1974, pages 86–112
57. Brylev G.B., Gashina S.B., Nizdoyminoga G.L. Radar characteristics of clouds and rainfall. - L.: Hydrometeoizdat, 1986. - 231 pages
58. Gorelik A.G., Chernikov A.A. Some results of a radar research of structure of the wind field at the heights of 50–700 m Works of the CAO, 1964, issue. 57. - p. 3-18
59. Ivanov A.A., Melnichuk Yu.V., Morgoyev A.K. The technique of assessment of vertical velocities of air movements in heavy cumulus clouds by means of the doppler radar. Works of the CAO, 1979, issue 135, pages 3–13
60. Stepanenko V. D. Radar-location in meteorology. - the 2nd ed. - L.: Hydrometeoizdat, 1973. - 343 pages
61. Atlas D. Achievements of radar meteorology / Translation from English - L.: Hydrometeoizdat, 1967. - 194 pages
62. Battan L. G. Radar meteorology. / Translation from English - L.: Hydrometeoizdat, 1962. - 196 pages
63. Bin B.R., Dutton E.D. Radio meteorology. - L.: Hydrometeoizdat, 1971. - 362 pages
64. Doviak R.J., Zrnich D.S., Sirmans D.S. Meteorological doppler radar stations: Review. TIIER, 1979, v. 67, No. 11, pages 63–102
65. Voskresensky V. P., Mikhaylutsa K.T., Chernulich V.V. State and the prospects of development of airborne meteo radar stations. Issues of special radio electronics, ser. RLT, 1981, issue 13, p. 85-92
66. Lhermitte R.M. Doppler radars as severe storms sensors. Bulletin of the American Meteorological Society, 1964, v. 45, p.p. 587–596
67. Melnikov V.M. Connection of average frequency of maxima of an output signal of the doppler radar with characteristics of the movement of lenses. Works of VGI, 1982, issue 51, p. 17–29
68. Ryzhkov A.V. Influence of inertia of hydrometeors on statistical characteristics of a radar signal. Works of GGO, 1982, issue 451, pages 49–54
69. Melnichuk Yu.V., Gorelik A.G. About connection of statistical properties a radio echo with the movement of lenses in clouds and rainfall. Works of the CAO, 1961, issue 36
70. Melnichuk Yu.V., Chernikov A.A. Operational method of detection of turbulence in clouds and rainfall. Works of the CAO, 1973, issue 110, p. 3–11
71. Melnik Yu. A. Some opportunities of use of the method of synthesizing of apertures for observation of meteorological objects. Works of GGO, 1978, issue 411, p. 107–112
72. Ryzhkov A.V. About a possibility of use of the method of the synthesized apertures in problems of radar meteorology. Works of GGO, 1982, issue 451, p. 107–118
73. Melnikov V.M. About definition of a spectrum of a meteoradio echo by means of measurement of frequency of emissions of an output signal of the radar. Works of VGI, 1982, issue 51, p. 17–29
74. RTCA DO-220. « Minimum Operational Performance Standards (MOPS) for Airborne Weather Radar with Forward-Looking Windshear Detection Capability » . - Radio Technical Commission for Aeronautics (RTCA), US, Washington, DC, September 21, 1993. Change 1 issued June 6, 1995
75. RTCA DO-173. « Minimum Operational Performance Standards for Airborne Weather and Ground Mapping Pulsed Radars » . - RTCA, US, Washington, DC, November, 1980
76. RTCA DO-178B. « Software Considerations in Airborne Systems and Equipment Certification ». - RTCA, US, Washington, DC, December 12, 1992. Errata issued March 26, 1999
77. Lewis M.S., Robinson P.A., Hinton D.A., Bowles R.L. The relationship of an integral wind shear hazard to aircraft performance limitations. NASA TM-109080
78. Radar stations with digital synthesizing of an antenna aperture. / V.N. Antipov, V.T. Goryainov, A.N. Kulin et al.; Eds. V.T. Goryainov. - M.: Radio and Communication, 1988. - 304 pages

79. Radar methods of a research of the Earth / Eds. Yu.A. Melnik. - M.: Sov. radio, 1980. - 264 pages
80. Bochkaryov A.M., Strukov Yu.P. The airborne radio-electronic equipment of aircraft. The results of science and technology. VINITI. Ser. Aircraft industry. 1990, v. 11, page 4–248
81. Multipurpose pulse-doppler radar stations for fighter planes: Review. - M.: RDE NIIAS, 1987. - 70 pages
82. Kanashchenkov A.I., Merkulov V.I., Samarin O.F. Shape of perspective airborne radar-tracking systems. Opportunities and restrictions. - M.: IPRZHR, 2002. - 176 pages

Chapter 2
The Mathematical Model of Detection of Meteorological Objects by Airborne RS and the Assessment of Danger for Air Vehicles Flights

2.1 The Structure of the Mathematical Model

The object of researches is an airborne RS working in the mode of detection and assessment of MO danger. The airborne RS together with a set of the objects is located within the vision area, and the carrier (AV) form the radar channel [1]. In the channel when carrying out R observation the process of transformation of the EMW field, reflected by objects, into the field of R signals which then are transformed to the R field of the targets [2] is carried out.

As algorithms of the solution of a problem of detection and assessment of MO danger are assumed by use of all main devices of a radar station, the organization of natural experiments by carrying out of numerous AV flights in various meteo-conditions and statistical processing of the received results demands big financial and other material resources. Semi-natural researches demand creation on the basis of real airborne RS of the corresponding land complex comparable to it on complexity and cost. Owing to the specified restrictions on organization of natural and semi-natural researches the main method of researches is carrying out of the imitating mathematical modelling representing reproduction on computer of a natural experiment with the use of the mathematical model describing the structure and functional connections of the modelled system, external influences on it, operation algorithms and rules of change of a condition of a system under the influence of external and internal perturbations. Modelling allows to consider the processes in a system practically at any level of specification and to realize various algorithms of its work. High performance of carrying out statistical experiments on modern computers, good reproducibility of initial conditions in the sets of model situations are the advantages of the method of imitating mathematical modelling.

Now several technologies of mathematical modelling are being developed [3, 4]. The most harmonious methodological working off characterizes the aggregate approach to creation of models [3] which is expedient to use for development of mathematical model of the considered task. In this case the model represents an aggregate system which is consistently broken into finite number of subsystems

Vereshchagin A. V. et al., *Signal Processing of Airborne Radar Stations*,
Springer Aerospace Technology, https://doi.org/10.1007/978-981-13-9988-6_2

with preservation of connections for ensuring their interaction. The structure of the aggregate model is synthesized on the known principles of block construction and specialization of mathematical models of radio systems [5]. Proceeding from the specified principles, the structure of the simulation model of detection and assessment of danger of MO zones should include three main parts (Fig. 2.1):

- the model of formation of R signals (MO model, model of dynamics of the RS carrier, model of a useful signal);
- the model of a path of processing of signals of airborne RS;
- the block of programs of control of an experiment and processing of results.

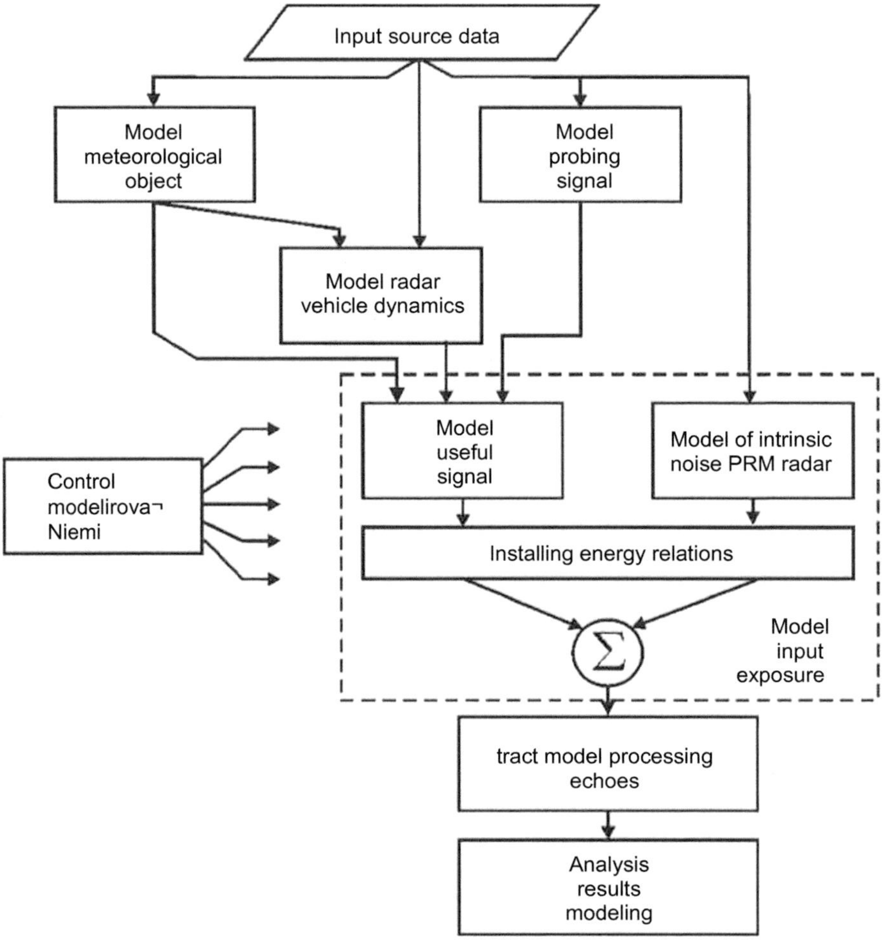

Fig. 2.1 The Structure of the mathematical model of the R channel at the solution of a problem of detection and assessment of danger of MO zones

When choosing the method of mathematical description of signals and interferences in the model it is necessary to consider that proceeding from the purpose and problems of an experimental study, the modelling of algorithms of digital processing of signals in an airborne RS should be carried out at the functional level. As the majority of modern devices of digital processing of signals (both reprogrammed and with unconditional logic) operates with digital readings of quadrature components of the complex envelope of the accepted signals, the method of complex envelope adequate to this representation is used in the developed model [4].

2.2 The Model of a Meteorological Object in the Conditions of Wind Shear and Turbulence

The target situation during the work of a radar station in the Turbulence and Wind shear modes is set by the MO type (cloud cover, fog, rainfall, etc.), its spatial sizes, structure (the structure of HM, their sizes, form and orientation), radar (specific EER, dielectric permeability, R reflectivity) and kinematic (a spectrum of velocities) characteristics.

MO (clouds, fogs, rainfall) as natural objects are extremely complex dynamic systems which are a subject to a large number of the influencing factors, such as wind phenomena, turbulence, temperature, pressure, solar radiation, etc., the majority of which has random character both in time and in space.

For the description of position and the movement of MO and other R targets in relation to a radar station we will introduce a number of systems of coordinates (Fig. 2.2):

1. The Cartesian system of coordinates connected with the RS carrier (the bound coordinate system, BCS) OXYZ. The O centre of BCS coincides with a RS APC. The direction of the axis OX coincides with the vector $\vec{W}$ of velocity of the RS carrier. The OZ axis is directed up along a local vertical, the OY axis is perpendicular to the OXZ plane.
2. The spherical system of coordinates connected with the RS carrier: range r, azimuth α and elevation angle β. The direction of the movement of the carrier in it is characterized by the angles $\alpha = \beta = 0$, the position of the axis of the RS ADP—by the angles α_0, β_0.
3. The Cartesian system of coordinates connected with the RS AS (or the antenna system of coordinates, ASC) $OX_aY_aZ_a$. The O centre of ASC coincides with APC. The OX_a axis is directed along axis of the RS ADP. The OY_aZ_a plane is combined with the aperture plane of the RS AS.
4. The spherical system of coordinates connected with the RS antenna system: the range r and the angles of sighting in the horizontal φ and vertical ψ planes. Let's orient this system of coordinates so that the plane $\psi(\varphi = 0)$ coincided with the elevation plane. Then the angles α, β and ψ, φ will be connected by the ratios

$$\varphi = (\alpha - \alpha_0)\cos\beta, \quad \psi = \beta - \beta_0. \tag{2.1}$$

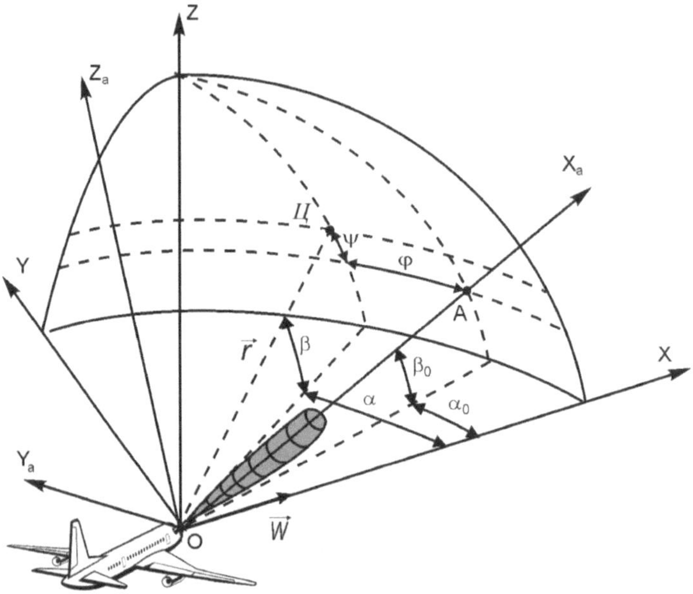

Fig. 2.2 The systems of coordinates used in the analysis of structure of the signals reflected by MO

2.2.1 The Parameters of a Meteorological Object Affecting the Efficiency of Radar Observation

In most cases the intensive descending flows are observed against the background of heavy cumulus (Annex 1). Therefore the key parameter defining efficiency of detection of the signals reflected by MO is its R Reflectivity (RR) which, in its turn, significantly depends on spatial distribution of MO water content. The value of MO water content depends on height in relation to MO lower bound [6]

$$\eta(\zeta) = \frac{\zeta^m (1-\zeta)^p}{\zeta_0^m (1-\zeta_0)^p} \eta\text{max}, \tag{2.2}$$

where $\zeta = z'/H$;

$z' = z - z_{\min}$	the height over the cloud basis;
$H = z\max - z_{\min}$	the MO thickness (power). Its average value $\bar{H} = 1800$ m [6];
m, p	the distribution parameters. The most often found (modal) values are equal to $\bar{m} = 2.8$ and $\bar{p} = 0.38$ [6];
ηmax	the maximum value of water content. It depends on the thickness of N and the temperature $T_{i\tilde{a}}$ at the cloud basis. At H = 2000 m and $T_{i\tilde{a}} = +10\,^\circ\text{C}$ the value of water content is 2.0 g/m^3 [6];

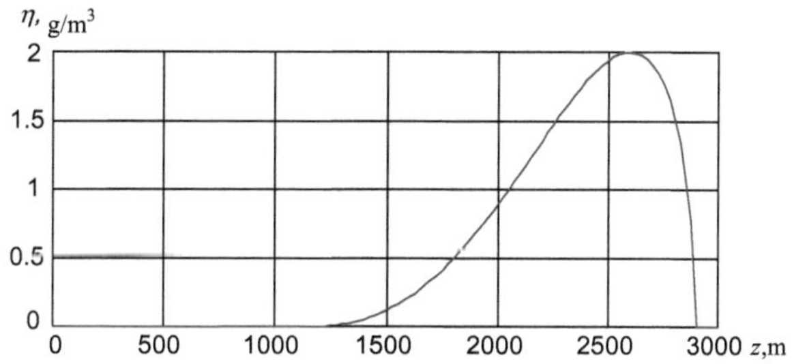

Fig. 2.3 Dependence of value of MO water content on height over the US level

ζ_0 the relative height of a layer of the maximum water content. Its most probable value is equal to $\bar{\zeta}_0 = 0.83 \pm 0.1$ [6];

z_{min} the height of the lower bound of cloud cover. In [7, page 180] on the basis of the analysis of large volume of these meteorological observations of cumulonimbus it is defined that z_{min} changes within a year within 0.8…1.3 km with the average value of 1.1 km

The Aquatic program (Annex 4) is specially developed for calculation in the MATLAB system of dependence of MO water content on height. The result of calculation is presented in Fig. 2.3.

Besides water content, the structure of the signal reflected by MO is also influenced by its other parameters. In particular, in the conditions of dangerous WS and atmospheric turbulence, as shown in Sect. 1.3, the spatial fields of average value and RMS width of a spectrum of wind speed are informative.

2.2.2 *The Model of a Meteorological Object in the Presence of Wind Shear*

On the basis of the analysis of a large number of the plane crashes which took place in 1970–80 it is found [8, 9] that at small height the intensive descending air flows arising at a stationary stage of development of storm cloud cover because of rain and the corresponding cooling of air are the main reason for formation of the heaviest WS. The typical descending flow represents a cylindrical vertical flow of cold air with a diameter of 1…4 km. At land impact air flow quickly spreads in many different directions (the Fig. 2.4a). The scientists T.T. Fujita and H.R. Byers suggested [10] to use the term "microburst" for designation of the descending air flows when a flow velocity is comparable or exceeds the value of velocity of an AV descent or ascent.

There are two simplified models of the wind field in the descending flow: the ring-vortex model [11] (the Fig. 2.4a) and the wall jet model [12, 13] (the Fig. 2.4b). According to the first model the descending air flow forms the three-dimensional axisymmetric vortex field in which the toroidal area ("core") where the wind speed, beginning from zero in the centre, linearly increases on radius to core border is

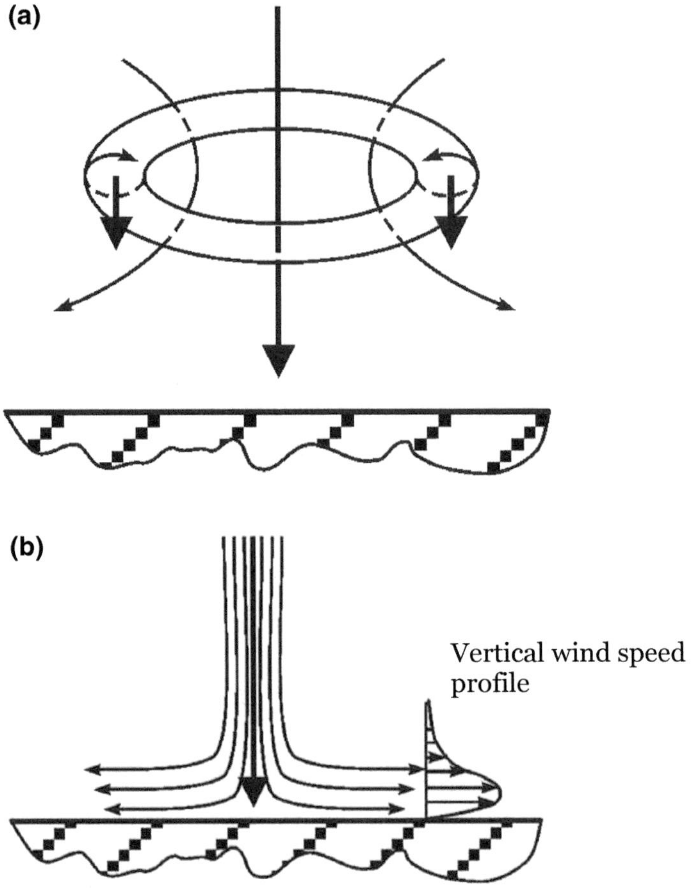

Fig. 2.4 The descending air flow (microburst): **a** the model of a ring vortex; **b** the wall jet model

located. This model is characteristic for the area of the falling air before it contacts with land. After land impact more accurate image of the movement of air is given by the second model according to which radial outflow of air in the form of near-surface jet takes place.

Let's further use the first of the presented models of the descending flow (Annex 2) as the problem of detection and assessment of degree of danger of the WS areas located at rather big height is considered.

The airborne RS in the course of observation determines the radial wind speed V_R in some point P (the centre of the allowed volume) as a wind speed projection to the direction connecting PAC to the specified point, (A2.18)

$$V_R = v_r \cos\left|\alpha_P - arctg\frac{y_V - y_P}{x_V - x_P}\alpha_P\right| \cos\beta_P + v_z \sin\beta_P, \tag{2.3}$$

where $\begin{cases} v_r = v\max\frac{z_P - z_V}{R_0}, \\ v_z = v\max\frac{r_P - r_V}{R_0}, \end{cases}$—the radial (A2.16) and vertical (A2.17) projections of wind speed in the point P;

$\begin{cases} x_V = r_V \cos\alpha_V \sin\beta_V, \\ y_V = r_V \sin\alpha_V \sin\beta_V, \\ z_V = r_V \cos\beta_V, \end{cases}$—the cartesian coordinates of the centre of a vortex in the BCS (A2.12);

$\begin{cases} x_P = r_P \cos\alpha_P \sin\beta_P, \\ y_P = r_P \sin\alpha_P \sin\beta_P, \\ z_P = r_P \cos\beta_P, \end{cases}$—the cartesian coordinates of the point P in the BCS (A2.13);

(r_V, α_V, β_V)—the polar coordinates of the vortex centre;

(r_P, α_P, β_P)—the polar coordinates of the point P;

$v\max = \frac{\nu}{2R_V}\left\{\left[1+\left(\frac{z_V - z_P}{R_V}\right)^2\right]^{-1,5} - \left[1+\left(\frac{-z_V - z_P}{R_V}\right)^2\right]^{-1,5}\right\}$—the maximum velocity on border of the vortex core (A2.15);

R_0—the vortex core radius;

R_V—the vortex radius in general;

$\nu = \frac{2V_V R_V}{1-\left[1+\left(\frac{2z_V}{R_V}\right)^2\right]^{-1,5}}$—the circulation of the vortex field (A2.7);

V_V—wind speed in the central part of a vortex flow.

Existence of a large number of the random factors influencing measurement of wind speed in a certain point of space in concrete timepoint owing to the central theorem of mathematical statistics leads to widening of a spectrum of wind speed. Thereof the ratio (2.3) is fair only for the description of the spatial field of average wind speed

$$\bar{V} = \frac{v\max}{R_0}\left[(z_P - z_V)\cos\left|\alpha_P - arctg\frac{y_V - y_P}{x_V - x_P}\alpha_P\right|\cos\beta_P + (r_P - r_V)\sin\beta_P\right] \tag{2.4}$$

The Radial program (Annex 4) is developed for calculation of an average radial wind speed at any point of MO in the MATLAB integrated programming system. The program contains a number of functions:

Stream the function of calculation of value of a ring vortex flow (A2.6);

Wind1 the function of calculation of projections of wind speed on (A2.8) and (A2.9);

Windmax the function of calculation of projections of wind speed on (A2.16) and (A2.17);

Wind velocity the function of calculation of average radial velocity of wind on (2.4)

2.2.3 The Model of a Meteorological Object in the Conditions of Intensive Turbulence

The uniform isotropic turbulent flow is completely characterized by the spatial correlation function (CF) or the corresponding spectral density and also the law of distribution of instantaneous values of wind speed. Spectral density of projections of the velocity of turbulent pulsations of wind to axes of a terrestrial system of coordinates are defined by the ratios [14, 15]

$$S_X(K_1) = 2\pi \int\limits_{K_1}^{\infty} F_X(\mathbf{K}) K dK, \; S_Y(K) = S_Z(K) = \frac{1}{2} S_X(K) - \frac{1}{2} \frac{d}{dK} S_X(K),$$

where **K**—the vector of spatial frequencies;
$F_X(\mathbf{K}) = \frac{1}{(2\pi)^3} \int R_X(\vec{\mathbf{r}}' - \vec{\mathbf{r}}) \exp(-j\mathbf{K}(\vec{\mathbf{r}}' - \vec{\mathbf{r}}))dV$—the tensor of spectral density [14] representing Fourier's transformation of CF of the uniform field of wind speed. For the isotropic vector field taking into account the continuity equation the tensor of spectral density has the form [16]

$$F_X(\mathbf{K}) = \left(1 - \frac{K_1^2}{K^2}\right) \frac{E(K)}{4\pi K^2}, \tag{2.5}$$

where K—the module of a vector of spatial frequencies;
K_1—the spatial frequency along the OX axis;
$E(K)$—the spectral density of an isotropic incompressible turbulent stream which at high spatial frequencies $(K \geq 10^{-3}\ \text{m}^{-1})$ satisfies the Kolmogorov-Obukhov law [17] $E(K) = 0.25\varepsilon^{2/3} K^{-5/3}$.

Thus, for the isotropic field one-dimensional spectra of projections of velocity of turbulent pulsations of wind are defined by the ratios [14]

$$S_X(K_1) = \frac{1}{2} \int\limits_{K_1}^{\infty} \left(1 - \frac{K_1^2}{K^2}\right) \frac{E(K)}{K} dK, \tag{2.6}$$

$$S_Y(K_1) = S_Z(K_1) = \frac{1}{4} \int\limits_{K_1}^{\infty} \left(1 + \frac{K_1^2}{K^2}\right) \frac{E(K)}{K} dK. \tag{2.7}$$

The CF of a projection of turbulent pulsations to the axis OX is generally described by the expression [15]

$$R_X(r) = \frac{(r/a)^n \sigma_X^2}{2^{n-1} \text{G}\,(n)} K_n(r/a), \tag{2.8}$$

$$R_Y(r) = R_Z(r) = \frac{(r/a)^n \sigma_{Y,Z}^2}{2^n \mathrm{G}\,(n)}\left[2K_n(r/a) - (r/a)K_{n-1}(r/a)\right], \tag{2.9}$$

where $K_n(x)$—the Bessel function an imaginary argument;
$\sigma_X^2, \sigma_Y^2, \sigma_Z^2$— the dispersions of turbulent fluctuations of wind speed;
n and a—the parameters defining a dependence form. The parameter a is connected with the external scale of turbulence with the ratio

$$a = L_0 \mathrm{G}(n) / \left[\sqrt{\pi}\mathrm{G}(n + 1/2)\right]. \tag{2.10}$$

Here G(n)—the gamma function.

From the expressions (2.8)–(2.10) it is possible to receive the ratios for spectral density a component of turbulent velocity in the form of [15]

$$S_X(K_1) = 4\sigma_X^2 L_0 \frac{1}{(1 + 4\pi^2 a^2 K_1^2)^{n+1/2}}, \tag{2.11}$$

$$S_Y(K_1) = 2\sigma_Y^2 L_0 \frac{1 + 8\pi^2 a^2 K_1^2 (n+1)}{(1 + 4\pi^2 a^2 K_1^2)^{n+3/2}}, \tag{2.12}$$

$$S_Z(K_1) = 2\sigma_Z^2 L_0 \frac{1 + 8\pi^2 a^2 K_1^2 (n+1)}{(1 + 4\pi^2 a^2 K_1^2)^{n+3/2}}. \tag{2.13}$$

The expressions (2.9)–(2.11) describe the dependences of turbulence energy distribution density on the scale L, dispersions of projections of turbulent speed and form parameter n. Changing the values of n, it is possible to receive a number of dependences of various form. In particular, two models [15, 18] are widely known. One of them is offered by Draydon. In it n = 1/2, then $a = L_0$ and

$$S_X(K_1) = 4\sigma_X^2 L_0 \frac{1}{1 + (2\pi K_1 L_0)^2}, \tag{2.14}$$

$$S_Y(K_1) = 2\sigma_Y^2 L_0 \frac{1 - 3(2\pi K_1 L_0)^2}{(1 + (2\pi K_1 L_0)^2)^2}, \tag{2.15}$$

$$S_Z(K_1) = 2\sigma_Z^2 L_0 \frac{1 - 3(2\pi K_1 L_0)^2}{(1 + (2\pi K_1 L_0)^2)^2}. \tag{2.16}$$

The corresponding correlation functions have the form

$$R_X(r) = \sigma_X^2 \exp(-r/L_0), \tag{2.17}$$

$$R_Y(r) = R_Z(r) = \sigma_{Y,Z}^2 \left(1 - \frac{1}{2}\frac{r}{L_0}\right)\exp\left(-\frac{r}{L_0}\right). \tag{2.18}$$

In the second model offered by Karman [19, page 30], n = 1/3, then

$$a = L_0 \frac{G(1/3)}{\sqrt{\pi} G(5/6)} \approx 1.339 L_0 \quad \text{and}$$

$$S_X(K_1) = 4\sigma_X^2 L_0 \frac{1}{\left[1 + (2\pi K_1 1.339 L_0)\right]^{5/6}}, \tag{2.19}$$

$$S_Y(K_1) = 2\sigma_Y^2 L_0 \frac{(1 + 8/3)(2\pi K_1 1.339 L_0)^2}{(1 + (2\pi K_1 1.339 L_0)^2)^{11/6}}, \tag{2.20}$$

$$S_Z(K_1) = 2\sigma_Z^2 L_0 \frac{(1 + 8/3)(2\pi K_1 1.339 L_0)^2}{(1 + (2\pi K_1 1.339 L_0)^2)^{11/6}}, \tag{2.21}$$

$$R_X(r) = 2^{2/3} \sigma_X^2 \frac{(r/a)^{1/3}}{G(1/3)} K_{1/3}(r/a), \tag{2.22}$$

$$R_Y(r) = R_Z(r) = 2^{2/3} \sigma_{Y,Z}^2 \frac{(r/a)^{1/3}}{G(1/3)} \left[K_{1/3}(r/a) - \frac{(r/a)}{2} K_{2/3}(r/a) \right], \tag{2.23}$$

where $K_{1/3}(x)$—the modified Bessel function of a fractional order of an imaginary argument.

The expressions obtained by Draydon are more convenient for use in analytical calculations, however the Karman model is more preferable at modelling as it is better coordinated with the theoretical description of vortexes in the atmosphere.

For modelling of random realization of turbulent fluctuations V_m of the radial wind speed estimated by an airborne RS in the course of Doppler measurements the spectral method can be used [20]. Thus after numerical calculation of a spectrum of turbulent fluctuations

$$S_X(K_1) = \int_0^\infty dr R_X(r) \exp(-j 2\pi K_1 r)$$

the realizations $V_m(r)$ are modelled in spectral area by means of the fast Fourier transformation (FFT)

$$V_m(i\delta r) = \operatorname{Re}\left\{ \sum_{k=0}^{N-1} \xi_k \left[\frac{1}{2N\delta r} S_X\left(\frac{k}{N\delta r} \right) \right]^{1/2} \exp\left(j 2\pi \frac{ki}{N} \right) \right\}, \tag{2.24}$$

where $i\delta r = r-$ the current value of range;
$i-$the number of an element of resolution on range;
$N-$the quantity of spectral channels (FFT points);
$\xi_k-$the complex pseudorandom value distributed under the normal law and satisfying to the following properties: $\langle \xi_k \xi_{k'} \rangle = 0$, $\langle \xi_k \xi_k^* \rangle = 1$.

Thus, as the first step ay formation of the R signals reflected by MO it is necessary to create the model of the meteorological object considering the spatial distribution of the MO characteristic parameters, in particular, the spatial structure of the three-dimensional field of radial wind speed in the conditions of shear and turbulence and also the spatial structure of the water content field.

2.3 The Mathematical Model of Movement of the Carrier of Airborne RS

The own movement of the RS carrier leads to additional movement of all R targets in a view zone in relation to RS APC. As the airborne RS in the course of Doppler measurements determines radial velocities of relative movement of R targets, the own movement of the AV—the RS carrier distorts the valid spectrum of radial velocities of R targets.

In [21] it is shown that the spatial movement of the RS carrier can be considered as superposition of the movement on a fixed basic trajectory (for example, uniform straight flight) and random AV deviations from a basic trajectory caused by a continuous random change of wind direction and speed, pressure and density of air, the module and the direction of an air speed vector, etc., and also its angular fluctuations. The AV control system, on the one hand, not always manages to react to these deviations and to support the set flight mode, and on the other—brings the elements of chance in the AV movement.

The random AV movement consists of the random AV movement as a solid body (trajectory irregularities, TI) and elastic displacements of PAC (elastic mode of a structure, EMS) [22]

$$\vec{r} = \vec{r}_{эк} + \vec{r}_{ук}\,, \tag{2.25}$$

where $\vec{r}_s$—the vector of the random AV movement as a solid body;
$\vec{r}_{ed} = \vec{r}_{cm} + \vec{r}_{вр}$—the vector of elastic displacement of PAC;
$\vec{r}$cm—the vector of the random movement of the centre of masses (CM) of elastic AV;
$\vec{r}_{вр} = \underline{\Psi}\,\vec{r}_a$—the vector of the random movement of PAC in relation to CM because of angular fluctuations of an AV on a course, tangage and rolling motion;
$\underline{\Psi} = \begin{vmatrix} 1 & -\psi_1(t) & \psi_2(t) \\ \psi_1(t) & 1 & -\psi_3(t) \\ -\psi_2(t) & \psi_3(t) & 1 \end{vmatrix}$—the matrix of transition from the BCS to the terrestrial system of coordinates at small values of angles of a course $\psi_1(t)$, tangage $\psi_2(t)$ and rolling motion $\psi_3(t)$;
$\vec{r}_a = \begin{vmatrix} x_a & y_a & z_a \end{vmatrix}^T$—the vector of the position of PAC in relation to CM in the BCS.

TI represent the reaction of an AV to influence of the turbulent atmosphere and noise of a control system. They include AV deviations from set flight trajectories, its angular fluctuations (heading drift in the azimuthal plane, emergence of an angle of attack in the vertical plane and rolling motion in relation to a construction axis), random changes of the module and the direction of a vector of speed, etc. On the other hand, under the influence of aerodynamic forces when flying in the turbulent atmosphere there are elastic modes of structural elements of AV—EMS which can be quite considerable at modern heavy high-speed AV. In real flight conditions TI and EMS, being present at the same time, cause random deviations of a trajectory of APC that leads to amplitude and phase distortions of the signals arriving to an input of the RS receiver. Such distortions are called trajectory distortions [21]. They can significantly worsen accuracy characteristics of the station.

For the assessment of the influence of TI and EMS on the efficiency of detection and determination of danger of WS and turbulence it is necessary to know their statistical characteristics. Their study by means of the system mathematical models of the AV movement in the turbulent atmosphere showed [22] that the densities of probability of any TI and EMS parameter can be approximated by normal stationary random process with zero mathematical expectation and CF of the type

$$K_i(\tau) = \sigma_i^2 e^{-\frac{\tau^2}{T_i^2}} \cos\frac{\pi\tau}{2\tau_i}, \tag{2.26}$$

where σ_i^2—the dispersion of the i-th parameter;
T_i—the interval of correlation of the i-th parameter;
τ_i—the conditional time of correlation determining the frequency of fluctuations of the correlation function. For all types of the AV random movement the condition $T_i \gg \tau_i$ [22] is satisfied, and for the intervals of observation $T_н \leq 1$ s the ratios $T_i >> \tau_i > T_н$ for TN and $T_i > T_н \geq \tau_i$ for EMS [21], and $T_i/\tau_i = 5\ldots10$ both for TI and for EMS [21] are observed [21].

The analysis carried out in [23] showed that TI is slow processes with the correlation interval from units to tens of seconds depending on the AV type. In particular, τ_i for TI changes within 2…12 s, for angular fluctuations—1…5 s. The RMS value σ_i of linear deviations for TI is units-tens of metres, angular deviations on a rolling motion and course—up to 1…2° (on tangage—by 3…5 times less). The own frequencies of AV EMS are much higher, than for TI and usually have several tones, the first of which is close to 1…2 Hz (τ_i is close to 1 s [21]). The value σ_i of linear deviations at EMS is in the range from the tenth shares of millimetre to several centimetres depending on flexibility of an AV structure.

At the majority of evolutions of an AV its movement is characterized by very small frequencies therefore in these conditions EMS have no significant effect on dynamics of flight owing to what VA is usually considered as a solid body. At the same time at flight in the turbulent atmosphere overloads change very sharply, practically repeating wind speed pulsations. In these conditions EMS can have a strong impact on dynamics of flight of an AV. At the installation of an AS in an AV fuselage the influence of EMS is especially strongly shown on heavy AV with a flexible structure. With reduction of the AV sizes and increase in rigidity of its structure the intensity of fluctuations of the fuselage decreases and for rather light AV of front aircraft they can be neglected. The most essential types of EMS are the bend and torsion of a wing and also the bend of a fuselage along a construction axis.

Aerodynamic vibration also has an impact on the position of AV structural elements. The character and intensity of vibration significantly depend on a mode of flight and a type of pilotage. At the transonic mode (0.8 M << W < 1.0 M) the amplitude of vibration is by 3…5 times more, than at the subsonic and supersonic modes of flight. The frequency range of fluctuations is within 3…1000 Hz. Total vibration accelerations in frequency range up to 300 Hz are 10 g and above. Vibroshifts of AV structural elements are maximum along the axes Y and Z of the BCS and amount to millimetres.

Thus, the resulting position of ARC in relation to a basic trajectory is defined by TI, the rotation of the fuselage around CM, flexural fluctuations and torsion of an AV structure and also aerodynamic vibration. All specified fluctuations have narrow-band character and are described by the CF of the type (2.26). In this case the dynamic model of an AV movement in the turbulent atmosphere can be presented by the system of the linear differential equations with constant coefficients for each of coordinates [18]. The equations of the system have the form

$$a_0 \frac{d^n y}{dt^n} + a_1 \frac{d^{n-1} y}{dt^{n-1}} + \cdots + a_n y = b_0 \frac{d^m x}{dt^m} + b_1 \frac{d^{m-1} x}{dt^{m-1}} + \cdots + b_m x, \tag{2.27}$$

where x—the input disturbance;
y—any coordinate of an AV;
a_i, b_j—the constant coefficients depending on an AV design.

The element with the transfer function (TF) of the following form corresponds to the linear differential equation (2.27)

$$H(j\omega) = \frac{b_0 (j\omega)^m + b_1 (j\omega)^{m-1} + \cdots + b_m}{a_0 (j\omega)^n + a_1 (j\omega)^{n-1} + \cdots + a_n}. \tag{2.28}$$

Thus, AV can be considered as a linear filter with TF (2.28) to which input the random signal in the form of random influence of the turbulent atmosphere arrives. Then the spectral density of fluctuations along coordinate y is equal to

$$S_y(\omega) = |H(j\omega)|^2 S_x(\omega), \tag{2.29}$$

where $S_x(\omega)$—spectral density of input disturbance.

Various analytical expressions for the TF of longitudinal and lateral motion of an AV at influence of wind are known [18, 24]. In particular, at the autopilot control profile [18]

$$\Delta\delta_{в} = -i_{\vartheta}(\Delta\vartheta_{з} - \Delta\vartheta) + i_{d\vartheta/dt}\frac{d\Delta\vartheta}{dt} + i_z\Delta\vartheta_{в}, \tag{2.30}$$

where $\Delta\delta_{в}$—the angle of elevator deflection;
i_{ϑ}—the gear ratio for the attitude angle;
$\Delta\vartheta_{з}$—the predefined increment value $\Delta\vartheta$ of the attitude angle;
$i_{d\vartheta/dt}$—the gear ratio for the angle rate;
i_z—the gear ratio for the height.

The transfer functions have the form:

- for the attitude angle at influence of a vertical wind component

$$H_{\vartheta z}(j2\pi f) = \frac{c_0(j2\pi f)^3 + c_1(j2\pi f)^2 + c_2(j2\pi f) + c_3}{(j2\pi f)^5 + d_1(j2\pi f)^4 + d_2(j2\pi f)^3 + d_3(j2\pi f)^2 + d_4(j2\pi f) + d_5};$$

- for the attitude angle at influence of a horizontal wind component

$$H_{\vartheta x}(j2\pi f) = \frac{e_0(j2\pi f)^3 + e_1(j2\pi f)^2 + e_2(j2\pi f)}{(j2\pi f)^5 + f_1(j2\pi f)^4 + f_2(j2\pi f)^3 + f_3(j2\pi f)^2 + f_4(j2\pi f) + f_5};$$

- for the banking angle at influence of a side wind component

$$H_{\gamma y}(j2\pi f) = \frac{1}{W}\frac{g_0(j2\pi f)^3 + g_1(j2\pi f)^2 + g_2(j2\pi f)}{(j2\pi f)^5 + p_1(j2\pi f)^4 + p_2(j2\pi f)^3 + p_3(j2\pi f)^2 + p_4(j2\pi f) + p_5};$$

- for the azimuth angle at influence of a side wind component

$$H_{y}(j2\pi f) = \frac{1}{W}\frac{q_0(j2\pi f)^3 + q_1(j2\pi f)^2 + q_2(j2\pi f)}{(j2\pi f)^5 + p_1(j2\pi f)^4 + p_2(j2\pi f)^3 + p_3(j2\pi f)^2 + p_4(j2\pi f) + p_5};$$

- for the CM coordinate y at influence of a side wind component

$$H_{\vartheta z}(j2\pi f) = \frac{h_0(j2\pi f)^4 + (k_1 - q_0)(j2\pi f)^3 + (k_2 - q_1)(j2\pi f)^2 + (k_3 - q_2)(j2\pi f) + l_5}{(j2\pi f)\left[(j2\pi f)^5 + l_1(j2\pi f)^4 + l_2(j2\pi f)^3 + l_3(j2\pi f)^2 + l_4(j2\pi f) + l_5\right]};$$

where W—the AV air velocity in not undisturbed atmosphere;
$c_i, d_i, e_i, f_i, g_i, h_i, k_i, l_i, p_i, q_i$—the constant coefficients depending on design data of an AV and the autopilot control profile.

In the absence of these results from natural flight experiments on the basis of which the coefficients $c_i, d_i, e_i, f_i, g_i, h_i, k_i, l_i, p_i, q_i$ are estimated, in the first approximation it is possible to consider an AV the linear system consisting of consistently connected oscillatory and inertial components with transfer function of the form (for the model No. 1 of a hypothetical AV [23])

$$H(p) = \frac{K_\delta}{p^2 + 4\pi d_K f_K p + (2\pi f_K)^2} \frac{1}{1 + T_V p},$$

where K_δ— the transfer coefficient on an angle (tangage, course). In the channel of stabilization of the position of an AV on tangage $K_\delta \approx V_z / W$;
d_K— the coefficient of damping of fluctuations (about 1–3);
f_K— the frequency of own short-period oscillations of an AV ($\approx$2 Hz);
T_V— the aerodynamic constant of time of an AV (about 2–3 s for Tu-154 M).

Having described the AV dynamic properties by means of the TF, it is possible, using (2.29), to receive spectral densities of fluctuations of the AV APC in relation to basic trajectory and spectral densities of values of angles of tangage, drift and rolling motion of an AV around APC. The instantaneous values of these parameters completely define the current position of the RS APC.

2.4 The Mathematical Model of a MO Radio Signal

2.4.1 The Structure of the Radio Signal Reflected by MO

The analysis of objectives demands the approach to MO as to spatially (volume) distributed R target consisting of a set of large number of randomly located independently moving elementary reflectors—hydrometeors (HM). Therefore the strict solution of a task of definition of MO R characteristics is not possible. Approximate methods of the solution of this task are built on the basis of the combination of provisions of electrodynamics and the theory of random processes.

Let EMW radiated by the RS AS fall to the area filled with the reflecting MO particles. Let's break all volume of MO into the separate allowed V_i volumes. In a distant zone it is possible to consider approximately that the volume of *Vi* has the cylinder form with the basis

$$r^2 \int_0^{2\pi} \int_0^{\pi} G^2(\alpha, \beta) \sin \alpha \, d\alpha d\beta \text{ and the height } \int_0^{\infty} w^2(r) dr.$$

Here $G(\alpha, \beta)$ is the two-dimensional AS DP on power, and weight function $w(r)$ describes the RS PS form on range [25, page 74]. At the DP Gaussian form we will receive

$$\int_0^{2\pi} \int_0^{\pi} G^2(\alpha, \beta) \sin \alpha d \alpha d \beta = \frac{\pi}{4} \frac{\Delta\alpha \Delta\beta}{2 \ln 2}.$$

The type of weight function on range $w(r)$ is determined by frequency characteristic of a RS reception path and a PS spectrum. If we introduce the L_r coefficient considering the extent of coordination of a PS spectrum with frequency characteristic of the RS receiver [14, page 78], then

$$\int_0^\infty w^2(r)dr = \frac{c\tau_{\grave{e}}}{2} L_r.$$

In case of pulse PS with rectangular envelope for Gaussian AFC and linear PFC of IFA filters of the RS receiver $L_r \approx 0.589$ [14, page 79].

Then the RS allowed volume is equal to

$$V = r^2 \int_0^{2\pi}\int_0^{\pi} G^2(\alpha,\beta)\sin\alpha d\alpha d\beta \int_0^\infty w^2(r)dr = \frac{\pi}{16\ln 2} r^2 \Delta\alpha\Delta\beta c\tau_{\grave{e}} L_r \tag{2.31}$$

Generally the repeated rereflection of EMW by hydrometeors takes place. However, as shown in [26], for pulse RS with an impulse duration $\tau_i \leq 1$ microsec it is possible not to consider recurrence of rereflections for the majority of MO. Then dispersion of PS by any allowed MO volume can be considered as superposition of partial signals of separate elementary reflectors mutually not influencing the processes of reflection of EMW by them.

The signal (1.6) reflected by a set of independently moving HM located in V_i volume contains the spectrum of Doppler frequencies corresponding to a spectrum of radial components of HM velocities. As mutual movement of reflectors leads to small, in comparison with the carrier frequency, widening of a spectrum, the signal (1.6) will represent narrow-band random process (1.12). The distribution of the amplitude factor of a signal (1.12) is connected with distribution of specific EER on the volume of MO and, as shown in the Sect. 1.2, is described by Rayleigh distribution, and the random phase of a signal (1.12) is evenly distributed on the interval $[-\pi, \pi]$ [27, 28].

As in modern and perspective airborne RS processing of the signals reflected by R targets is made by digital methods at a video frequency, the signal (1.12) after linear analog processing and quadrature phase detecting in the RS receiver is transformed in the ADC block to discrete readings of quadrature components of the complex envelope [29]

$$\dot{S}(t) = A(t)\exp[j\varphi_0 + j\varphi(t)]. \tag{2.32}$$

For coherent and pulse airborne RS with LRF and ARF of PS the discrete readings of the complex envelope of the signals reflected by any allowed MO volume at an input of PSP can be presented in the form of the vector

$$\vec{\mathbf{S}} = [\dot{S}_m], m = \overline{1, M}, \tag{2.33}$$

where $\dot{S}_m = \dot{S}(t_m) = \dot{S}(mT_n) = \dot{S}^c_m + j\dot{S}^s_m$—the vector elements (2.36) representing additive mixture of the complex signal envelopes of all elementary reflectors in this allowed volume

$$\dot{S}_m = \sum_n \dot{S}_n(mT_n) = \sum_n A_n G(\alpha_n) G(\beta_n)\sigma_n^{1/2} \exp\left[j2\pi f_0 \pm j4\pi \frac{V_n(mT_n)}{c} f_0 m T_n \right]; \quad (2.34)$$

where M—the number of impulses in the sequence of the reflected signals (the pack size);
m—the number of an impulse in a pack;
$\dot{S}^c_m = A(mT_n)\cos[\varphi_0 + \varphi(mT_n)]$ and $\dot{S}^s_m = A(mT_n)\sin[\varphi_0 + \varphi(mT_n)]$—cosine (in-phase) and sinus (quadrature) components of the complex MO signal envelope.

Thus, the problem of creation of the model of a MO signal comes down to the problem of creation of the mathematical model of two statistically connected discrete random processes representing two quadratures of the complex envelope of the signal reflected by the allowed MO volume.

Let the radial component of average wind speed $\bar{V} = 0$. Let the radial component of average wind speed = 0.

If within the allowed volume there is rather a large number of reflectors, then on the basis of the central limit theorem of probabilities the specified quadrature components of complex envelope of the signal reflected by the allowed volume of MO can be considered as mutually independent Gaussian random processes [23, 30, 31] with zero mathematical expectation and the dispersion σ_s^2, equal to the average power of the reflected signal

$$\dot{S}^c_{m0} = A(mT_n)\cos\,\varphi_0 \text{ and } \dot{S}^s_{m0} = A(mT_n)\sin\,\varphi_0.$$

The amplitudes of quadrature components are distributed under Rayleigh law, and phases are evenly distributed on the interval $[-\pi, \pi]$.

Now we will consider the case when $\bar{V} \neq 0$, at the same time the shift of the spectrum of the reflected signal on $\bar{\omega}$ takes place. The quadrature components of complex signal envelope are defined by the expressions

$$\begin{aligned}\dot{S}^c_m &= A(mT_n)\cos[\varphi_0 + \varphi(mT_n)] = A(mT_n)\{\cos\,\varphi_0 \cos\,\varphi(mT_n) - \sin\,\varphi_0 \sin\,\varphi(mT_n)\}\\ &= \dot{S}^c_{m0}\cos\,\varphi(mT_n) - \dot{S}^s_{m0}\sin\,\varphi(mT_n),\end{aligned}$$

$$\begin{aligned}\dot{S}^s_m &= A(mT_n)\sin[\varphi_0 + \varphi(mT_n)] = A(mT_n)\{\sin\,\varphi_0 \cos\,\varphi(mT_n) + \cos\,\varphi_0 \sin\,\varphi(mT_n)\}\\ &= \dot{S}^s_{m0}\cos\,\varphi(mT_n) + \dot{S}^s_{m0}\sin\,\varphi(mT_n),\end{aligned}$$

are linear combinations of normal random processes $\dot{S}^c_{m0}$ $\dot{S}^s_{m0}$ and represent normal processes.

After averaging on a set of realization we will receive that mathematical expectations of random processes $\dot{S}^c_m$ and $\dot{S}^s_m$ are equal to zero, i.e. they will coincide with mathematical expectations of the processes $\dot{S}^c_{m0}$ and $\dot{S}^s_{m0}$, and the dispersions are equal to σ_s^2, that is coherent with the physical side of the process as in the presence of wind energy of the reflected signal, on average, does not change.

Thus, in the absence of wind the mathematical model of the signal reflected by the allowed MO volume represents the model of two independent Markov processes of the first order [32, 33]. At presence of wind the quadrature components of the complex envelope of the signal become the correlated Markov processes of the second order.

For a complete statistical description of the signal reflected by the allowed MO volume the knowledge of only one-dimensional law of distribution is not enough. It is also necessary to estimate the connection between its various instantaneous values characterized by complex space-time generally non-stationary ACF

$$B(t_1, t_2) = \overline{s(t_1)s^*(t_2)}.$$

Let us set $\tau = t_2 - t_1; t = (t_2 + t_1)\big/2$, then

$$B(\tau, t) = \overline{s(t - \tau\big/2)s^*(t + \tau\big/2)}. \tag{2.35}$$

Though the expression (2.35) includes full information on power and spectral properties of the reflected signals, it is convenient to divide these data, having written down ACF in the form of the product

$$B(\tau, t) = P_c(t)\rho(\tau, t),$$

where $P_c(t) = B(0, t) = \overline{s(t)s^*(t)} = \overline{|s(t)|^2}$—the power of a signal (1.12); $\rho(\tau, t) = \frac{B(\tau,t)}{P_c(t)} = \frac{\overline{s(t-\tau/2)s^*(t+\tau/2)}}{\overline{s(t)s^*(t)}}$—the rated ACF (or correlation coefficient) of a signal (1.12).

As shown in [34], the rated ACF of fluctuations of the reflected signals can be presented as the product of two factors

$$\rho(\tau, t) = \rho_r(\tau, t)\rho_g(\tau, t). \tag{2.36}$$

The first cofactor describes the slow fluctuations caused by difference of Doppler frequencies of elementary signals within the allowed volume (interperiod fluctuations). The second factor, being a periodic function τ, characterizes the fast fluctuations connected with distribution of periodic PS in space (intra period fluctuations).

As analytical determination of two-dimensional density of probability of signals (processes) is difficult for real MO [26, 35], and numerical modelling owing to the small accuracy of setting of characteristics of a MO microstructure gives the result with a big error, usually the rated ACF is defined experimentally.

The rated temporal ACF $\rho_r(\tau, t)$ of a signal of the allowed volume of stratocumulus cloud cover [30, page 248] characterizing interperiod fluctuations of a signal is given as an example in Fig. 2.5, and in Fig. 2.6—the rated spatial ACF $\rho_g(\tau, t)$,

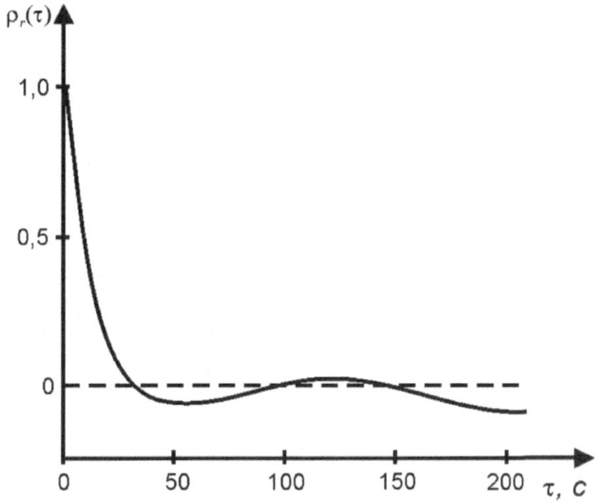

Fig. 2.5 The temporal rated ACF of the reflected signals for stratocumulus

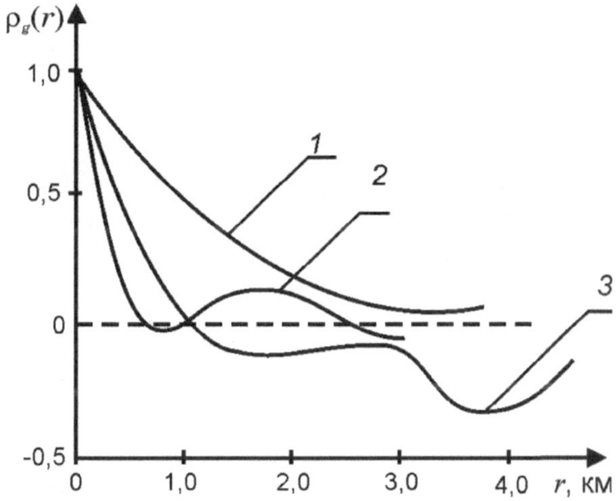

Fig. 2.6 The spatial rated ACF of the reflected signals: 1—for stratocumulus; 2—for stratus; 3—for cumulonimbus

characterizing variability of intensity of the reflected signals in space from one volume to another [26, 35].

The analysis of ACF $\rho_r(\tau, t)$ (Fig. 2.5) shows the following:

(a) ACF of the signals reflected by stratocumulus is almost equal to the unit for the intervals of time exceeding 3 s;
(b) ACF of the signals reflected by stratocumulus for the intervals of time more than 60–90 s, is almost equal to zero, i.e. this value can be accepted as time of correlation of the specified signals [31, page 248].

The spatial radius of correlation which can be accepted as the first transition of the dependences given in Fig. 2.6 through zero level changes from 0.7 to 3.3 km depending on a cloud cover type. It should be also noted that correlation properties of the reflected signals significantly depend on the range to MO [36]. So, for signals with $\lambda = 3$ cm the reduction of the interval of correlation τ at increase in range from 5 to 100 km can be approximated by the formula

$$\tau(r) = \tau_0 \exp(-\gamma r),$$

where τ_0—the correlation interval for initial range r_0; $\gamma = 0.021$.

The presented dependence can be explained with increase in the allowed V_i volume with increase of range. In this regard within V_i the reflectors with wide spacing of velocities are observed that leads to the DS widening and reduction of an interval of correlation of the reflected signals.

It is convenient to present the ACF values of interperiod fluctuations determined by mathematical expectation

$$\rho_r(t_i, t_k) = \frac{\overline{\dot{S}(t_i)\dot{S}^*(t_k)}}{\overline{\dot{S}(t_i)\dot{S}^*(t_i)}} = \frac{\overline{\dot{S}(t_i)\dot{S}^*(t_k)}}{\bar{P}_c},$$

in the form of the autocorrelated matrix (ACM) with the size MxM

$$\underline{\mathbf{B}}_r = [\rho_{r\,ik}], \tag{2.37}$$

which elements are set by the known ratios [37]

$$\rho_{r\,ik} = \rho_{0\,ik} \exp\left[-j(i-k)2\pi \bar{f} T_n\right], \tag{2.38}$$

$$\rho_{0\,ik} = \exp\left[-\frac{(i-k)}{2}\pi \Delta f^2 T_n^2\right]. \tag{2.39}$$

The instantaneous (time-and-frequency) spectrum of fluctuations

$$S(f,t) = \int_{-\infty}^{\infty} B(\tau, t) \exp(-j2\pi f \tau) d\tau, \tag{2.40}$$

being the generalized characteristic of the reflected signals in frequency area is connected with ACF of the reflected signals by means of Fourier transformation. Two spectra also correspond to two components of ACF (2.36): the spectrum of slow (Doppler, interperiod) fluctuations and the spectrum of fast fluctuations (intra period fluctuations, fluctuations on range). Let's consider further the first of them.

2.4.2 The Power Characteristics of the Radio Signal Reflected by MO

According to the equation of R observation range of spatially distributed targets filling all allowed volume, the average power of the signal (1.12) reflected by the allowed MO volume is equal to

$$\overline{P}_c = \int\limits_{V_\text{i}} P_\text{d} dV_\text{i} = \frac{P_\text{i} K_g^2 \lambda^2}{(4\pi)^3 r^4} F^4 \int\limits_{V_\text{i}} \sigma(V) dV_\text{i}, \tag{2.41}$$

where P_d—the specific power caused by dispersion of EMW by a unit of volume;
P_i—the power of the probing pulse signal of a RS;
K_g—the antenna amplification factor;
F—the coefficient of attenuation of intensity of the EMW field in the atmosphere;
$\sigma(V)$—the distribution of EER of elementary reflectors on MO volume.

The value of total EER of the allowed volume $\sigma_\Sigma = \int\limits_{Vi} \sigma(V) dV_\text{i}$ significantly depends on the ratio of coherent and incoherent components of dispersion of the falling EMW. In [26, 35] it is shown that the coherent dispersion can be neglected in the considered 3 cm wave band. In this case the total EER can be presented in the form of the product of the specific EER $\sigma_{\acute{o}\ddot{a}}$, characterizing intensity of reflections from MO and the equal to the sum of EER of elementary reflectors in MO unit of volume, with the value of the allowed volume

$$\sigma_\Sigma = \int\limits_{V_\text{i}} \sigma(V) dVi = \sigma_d V_\text{i}, \tag{2.42}$$

where

$$\sigma_d = \frac{\sigma_\Sigma}{V_i} = \frac{1}{V_i} \sum_{i=1}^{N} \sigma_i; \tag{2.43}$$

N—the number of elementary reflectors in the allowed volume.

Water drops, ice crystals, snowflakes, hailstones and snowgrains belong to MO particles reflecting PS. R characteristics of similar particles can be calculated by one of two known methods [26, 35]: the strict (according to the theory of diffraction of Mi) and the approximate (Rayleigh method), and the approximate method is a limit case of an accurate one at the small size of a particle in comparison with length of the falling EMW. As in the 3 cm range the condition of applicability of the Rayleigh method is met, EER of a spherical particle characterizing the amount of the energy disseminated by a particle in the direction of radar station is equal to [6, 38]

$$\sigma_i = 64\pi^5 \left| \frac{m^2 - 1}{m^2 + 2} \right|^2 \frac{R_i^6}{\lambda^4} = \pi^5 K_{\text{hm}}^2 \frac{D_i^6}{\lambda^4}, \tag{2.44}$$

where R_i and D_i—the radius and diameter of the scattering particle respectively; m—the complex index of refraction of a particle connected with its dielectric permeability ε by the ratio $m = \sqrt{\varepsilon}$;
$K_{\text{hm}} = \left| \frac{m^2-1}{m^2+2} \right|$—the coefficient characterizing dielectric properties of HM: for water $K_{\text{hm}}^2 = 0.93$ [39, 40], for ice $K_{\text{hm}}^2 = 0.197$ [31].

Substituting (2.44) in (2.41), we receive the following expression for specific the MO EER

$$\sigma_{\text{d}} = \frac{1}{V_i} \sum_{i=1}^{N} \frac{\pi^5 K_{\text{hm}}^2}{\lambda^4} D_i^6 = \frac{\pi^5 K_{\text{hm}}^2}{\lambda^4} \sum_{i=1}^{N} \frac{D_i^6}{Vi}.$$

The value $\sum_{i=1}^{N} \frac{D_i^6}{V_{\text{Ц}}}$ is called the R Reflectivity (RR) Z MO [41, page 5].

RR can be interpreted as the average sum of the diameters of particles in a unit of MO volume raised to the sixth power. In a general view, RR is defined by the function of distribution of MO particles on the sizes in a unit of volume $N(D)$

$$Z = \int_{0}^{+\infty} D^6 N(D) dD, \tag{2.45}$$

where $N(D)dD$—the number of HM with the diameter from D to $D + dD$, which are in the single MO volume. In practice RR of clouds and rainfall are measured in mm^6/m^3 or in dBz in relation to $Z_0 = 1\ \text{mm}^6/\text{m}^3$.

Taking into account the introduction of the RR concept the expression for the specific MO EER can be written down as

$$\sigma_{\text{d}} = \left(\pi^5 K_{\text{hm}}^2 / \lambda^4\right) Z. \tag{2.46}$$

Thus, the specific MO ERR is defined by the wavelength of a radar station and MO radar reflectivity.

Substituting (2.41) the expressions (2.42), (2.46) and (2.31) in the equation of R observation range, we will receive the expression for the average power of the signal reflected by the allowed MO volume,

$$\bar{P}_c = \frac{\pi^3}{2^{10} \ln 2} \frac{P_{\text{i}} G^2 \lambda^2 F^4 \Delta\alpha \Delta\beta c \tau_{\text{i}}}{\lambda^4} K_{\text{hm}}^2 \frac{Z}{r^2} L_r, \tag{2.47}$$

which taking into account the concept of energy potential (1.24) introduced earlier Π_M (1.24) can be written down in the form

$$\overline{P}_c = \Pi_{м} P_{\min} K_{\text{hm}}^2 \frac{Z}{r^2} L_r. \tag{2.48}$$

For determination of power parameters of the signal reflected by MO it is necessary to consider spatial distribution of its RR. Let's substitute in the expression (2.48) which gained the greatest distribution in the works on meteorological radar-location, the distribution of MO particles by the sizes in a unit of volume offered by Marshall and Palmer [38, 41] on the basis of approximation of empirical data

$$N(D) = \begin{cases} N_0 \exp[-\Lambda D] & at\ D \le 6\,\text{mm}, \\ 0 & at\ D > 6\,\text{mm}, \end{cases} \tag{2.49}$$

where $N_0 = 0.08\ \text{cm}^{-4}$—the normalizing parameter;
$\Lambda = 41\eta^{0.21}$—the parameter depending on water content η of MO (the Sect. 2.2).

Then the connection between RR and water content of MO will be defined by the power function of the form [31, 41]

$$Z = A\eta^b, \tag{2.50}$$

which parameters A and b depend on the MO type.

In particular, for powerful cumulus clouds $A = 16.3$, $b = 1.46$ [31], for nimbostratus clouds $A = 1380$, $b = 1.07$ [41]. In (2.50) the RR Z is measured in mm^6/m^3, the water content η—in g/m^3.

Taking into account (2.2) the spatial distribution of RR can be presented by the ratio

$$Z(\zeta) = A\left[\frac{\zeta^m(1-\zeta)^p}{\zeta_0^m(1-\zeta_0)^p}\eta_{\max}\right]^b. \tag{2.51}$$

The Reflectivity program (Annex 4) using the Aquatic procedure described in the Sect. 2.2 for calculation of water content of MO as the function (2.2) of height is developed in the MATLAB system for calculation of dependence of RR of the allowed MO volume from height. The results of calculation of RR by means of the Reflectivity program presented in the Fig. 2.7 show that RR significantly depends on height, and the greatest RR characterizes the top layers of MO.

Spatial distribution of the specific MO EER (2.46) for RR is defined taking into account the expression (2.51) by the ratio

$$\sigma_d(z) = \frac{\pi^5 \times 10^{-15} K^2}{\lambda^4} A\left[\frac{(z-z_{\min})^m (z_{\max}-z)^p}{(z_{\max}-z_{\min})^{m+p}} \frac{(m+p)^{m+p}}{m^m p^p}\eta_{\max}\right]^b. \tag{2.52}$$

In case the DP covers all MO angular size on height, the height of the centre of the allowed volume $z_0 = (z_{\max} + z_{\min})/2$, and the formula (2.52) becomes simpler

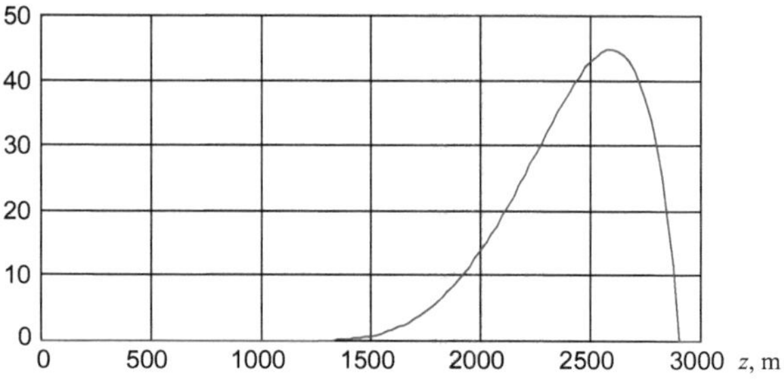

Fig. 2.7 Dependence MO of radar reflectivity on the height over the US level

$$\sigma_{\mathrm{d}} = \frac{\pi^5 \times 10^{-15} K^2}{\lambda^4} A \left[\frac{(m+p)^{m+p}}{2^{m+p} m^m p^p} \eta \mathrm{max} \right]^b. \tag{2.53}$$

The total EER of the allowed MO volume (2.52), taking into account (2.31) and (2.53), can be presented in the form

$$\sigma_{\Sigma}(r) = \frac{\pi^6 \times 10^{-15}}{16 \ln 2} \frac{r^2 \Delta\alpha \Delta\beta c \tau_{\grave{e}}}{\lambda^4} K^2 L_r A \left[\frac{(m+p)^{m+p}}{2^{m+p} m^m p^p} \eta \mathrm{max} \right]^b. \tag{2.54}$$

In Fig. 2.8 the dependence $\sigma_{\Sigma}(r)$ of total EER of the allowed MO volume on range received by means of developed in the MATLAB system SEpr procedure (Annex 4) realizing the ratio (2.54) for basic data is presented: $\Delta\alpha = \Delta\beta = 1°$, $\tau_u = 1$ μs, $K^2 = 0.93$, $\lambda = 0.03$ m, $L_r = 0.589$, $A = 16.3$, $b = 1.46$, $\eta\mathrm{max} = 2.0\ \mathrm{g/m^3}$, $m = 2.8$, $p = 0.38$. The carried-out calculation showed that MO EER for the specified parameters is very considerable (tens-hundreds of sq.m) and increases under the square law with increase in range (2.57). However at certain values of range owing to significant increase in the cross amounts of the allowed volume the condition of its filling with hydrometeors is violated (see Annex 5). At the same time the EER growth rate (in process of increase in range) decreases. When at some range the allowed volume covers the whole MO, the increase of EER will stop.

Spatial distribution of average power (2.51) of the signal reflected by the allowed MO volume, taking into account (2.54) is brought to the form

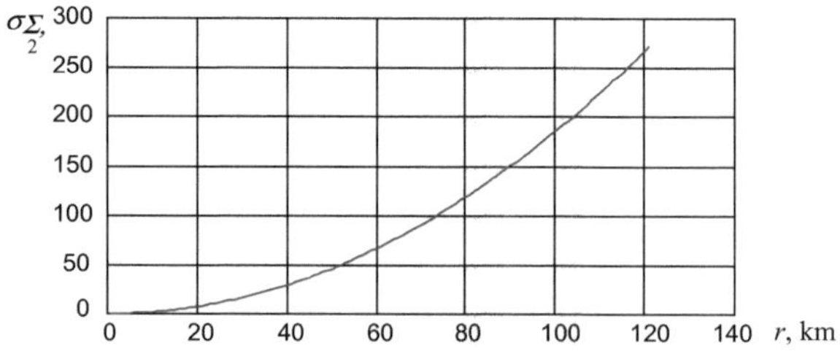

Fig. 2.8 The dependence of total EER of the allowed MO volume on range

$$\bar{P}_c(r,\beta) = \frac{\Pi_M P_{\min} K^2 L_r A}{r^2}\left[\frac{(z - z_{\min})^m (z_{\max} - z)^p}{(z_{\max} - z_{\min})^{m+p}} \frac{(m+p)^{m+p}}{m^m p^p} \eta_{\max}\right]^b. \tag{2.55}$$

In case DP covers the whole MO angular size on height, then by analogy with (2.53) the formula (2.55) also becomes simpler

$$\bar{P}_c(r_0) = \frac{\Pi_M P_{\min} K^2 L_r A}{r_0^2}\left[\frac{(m+p)^{m+p}}{2^{m+p} m^m p^p} \eta_{\max}\right]^b. \tag{2.56}$$

Here it should be noted that for ensuring constancy of characteristics of detection of the signals reflected by MO with identical EER in various ranges it is necessary to change as appropriate the energy (meteorological) potential of radar station, in particular—the pulse power of PS or its duration. However, increase in duration of PS significantly reduces the accuracy of assessment of danger of atmospheric turbulence (the Sect. 2.4.3.2).

Despite the difference in circuit decisions and element base, the vast majority of the existing domestic airborne RS carry out the RR assessment on the basis of the analysis of average power of the reflected signal (2.51)

$$Z = r^2 C_1^{-1} \overline{P}_c, \tag{2.57}$$

where $C_1 = \Pi_M P_{\min} K^2 L_r$—the coefficient considering power opportunities of a RS.

Generally average for the period of carrier frequency power of the signal reflected from the allowed MO volume is equal to [14]

$$\begin{aligned}\bar{P}_c &= B(0) = \overline{s(t)s^*(t)} \sim \sum_{n-1}^{N} \left(A_n G(\alpha_n) G(\beta_n) \sigma_n^{1/2}\right)^2 \\ &+ \sum_{n=1}^{N} \sum_{\substack{l=1 \\ l \neq n}}^{N} \left(A_n G(\alpha_n) G(\beta_n) \sigma_n^{1/2}\right)\left(A_l G(\alpha_l) G(\beta_l) \sigma_l^{1/2}\right) \exp[j 4\pi (r_n - r_l)/\lambda].\end{aligned} \tag{2.58}$$

The first summand in (2.58) is the constant not depending on position of HM within the allowed volume characterizes the average value $\bar{P}_c$ of power of the reflected signals, and, therefore, determines the RR value. The second summand represents the fluctuating part of instantaneous power depending on a relative positioning of reflectors in V_i and also from their position in relation to a radar station.

In spite of the fact that second summand in (2.58) in some cases can be much more than the first one (it contains N(N-1) components, whereas the first summand—only N components), at averaging on many consecutive readings it tends to zero (since the temporal average of an exponential multiplier in it tends to zero [14]). Therefore the exact assessment of the average power of the reflected signals characterizing RR is impossible without averaging of a signal on M to consecutive readings. At the same time

$$\overline{P}_c = \frac{1}{M}\sum_{m=1}^{M} P_{c\,m}, \tag{2.59}$$

where $P_{c\,m}$—the instantaneous power of the reflected signals in the m—the sounding period.

The total number M of readings of the signal reflected by the allowed volume is defined by time of its radiation. If the readings used in the course of obtaining of average value $\bar{P}_c$, are not correlated, then RMS assessment error $\bar{P}_c$, and consequently, and RR assessment at coherent accumulation is in proportion to 1/M. However, generally because of correlation of the reflected signals the mathematical expectation of the fluctuating part of the reflected signals power (2.58) becomes not equal to zero that increases the RMS RR assessment error. The degree of correlation depends both on the RS parameters (the period of repetition of impulses, the width of DP, sight angles of a beam), and on the MO parameters (turbulence degree, WS value). Besides, in case of placement of radar station on a board of a mobile carrier the degree of correlation also depends on the velocity of the movement of the carrier.

It is necessary to define the number of uncorrelated (effective) readings M_E. At equidistant readings the value of M_E is equal to [42]

$$\frac{1}{M_Э} = \frac{1}{M} + \frac{2}{M}\sum_{m=1}^{M-1}\left(1 - \frac{m}{M}\right)\rho(mT_n) \tag{2.60}$$

It follows from the expression (2.60) that if readings are not correlated through the interval of time T_p, then $\rho(mT_n) = 0$ for any t $\neq$ 0, and the expression (2.60) will turn into the equality $M_E = M$, that is the whole accumulated pack of the reflected impulses will be effective. But if readings are partially correlated, $M_Э < M$ and RMS error of RR assessment will increase though the size of a pack will remain equal to M readings.

2.4.3 The Spectral Characteristics of the Radio Signal Reflected by MO in the Conditions of Wind Shear and Turbulence of the Atmosphere

The R signal reflected by the allowed MO volume as it was noted in the Sect. 2.1, represents superposition of reflections from the particles filling this volume and contains the spectrum of frequencies corresponding to the spectrum of the radial components of velocities of separate reflectors. Here it should be noted that the DS of the reflected signals describes the distribution of radial velocities of reflectors taking into account the contribution of each of them to the reflected signal [25, 43] which is defined by the part of power of a signal of the allowed volume which contains in a partial signal of this elementary reflector.

2.4.3.1 The Spatial Distribution of Average Frequency and Width of a Doppler Spectrum of the Signal Reflected by MO

Let us consider a set of the reflectors setting in space the field of radial velocities $V(\vec{r}_n, t)$ and the RR field characterized by the distribution of specific EER $\sigma(\vec{r}_n, t)$, where $\vec{r}_n$—the radius vector of a particle of MO (Fig. 2.9). Let us assume that the fields $V(\vec{r}_n, t)$ and $\sigma(\vec{r}_n, t)$ are stationary, in any case, during observation, i.e. the values $V(\vec{r}_n, t)$ and $\sigma(\vec{r}_n, t)$ do not change during the specified time. Thus, $V(\vec{r}_n, t) = V(\vec{r}_n)$, $\sigma(\vec{r}_n, t) = \sigma(\vec{r}_n)$. Let the centre of the allowed volume will be located in the point $\vec{r}_0$, and the corresponding function of radiation for this volume is described by the expression

$$I(\vec{r}_0, \vec{r}_n) = \frac{C_1 G^4(\alpha_n - \alpha_0, \beta_n - \beta_0)|w(r_0, r_n)|^2}{r_n^4} \tag{2.61}$$

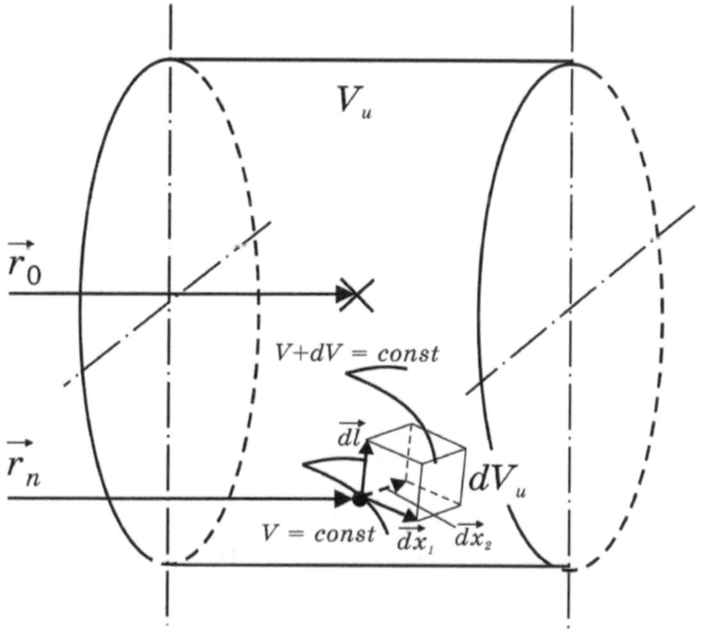

Fig. 2.9 The spatial ratios defining the spectrum of power of the signal reflected by MO

and id defined by AS DP $G(\alpha, \beta)$ and weight function on the range $w(r)$ (the Sect. 2.4.1). C_1—the constant determined by power opportunities of a RS.

Let us set in space an isotopic surface $V(\vec{r}_n) = const$ and find a total contribution to power from the reflectors moving with velocities from V to $V + dV$. The value of this contribution will be equal to the total power disseminated by the volume limited to two surfaces of constant velocities: V and $V + dV$. Let us set now the elementary volume, where dx_1 and dx_2—are the lengths of two orthogonal arches belonging to the surface $V(\vec{r}_n) = const$ which are crossed in the point $\vec{r}_n$ (Fig. 2.9).

The third coordinate dl is perpendicular to the surface $V(\vec{r}_n) = const$, can be determined by the expression

$$dl = |\text{grad}\, V(\vec{r}_n)|^{-1} dV.$$

The contribution to average power of the reflected signal from this elementary volume is equal to

$$d\bar{P}(V) = \sigma(\vec{r}_n) I(\vec{r}_0, \vec{r}_n) dx_1 dx_2 dl = \sigma(\vec{r}_n) I(\vec{r}_0, \vec{r}_n) |\text{grad}\, V(\vec{r}_n)|^{-1} dx_1 dx_2 dV. \tag{2.62}$$

The integration on an isotopic surface gives the value of total power for the interval of velocities $[V, V + dV]$, equal by definition to the product of density of a spectrum of power with dV

$$\bar{P}(\vec{r}_0, V) = \bar{S}(\vec{r}_0, V)dV = \left[\iint \sigma(\vec{r}_n) I(\vec{r}_0, \vec{r}_n) |\text{grad}\, V(\vec{r}_n)|^{-1} dx_1 dx_2\right] dV, \tag{2.63}$$

where $\bar{S}(\vec{r}_0, V)$—the average density of a spectrum of power.

Let us consider now the average radial velocity which value, as shown in [25], is weighed by distribution of power

$$\bar{V}_M(\vec{r}_0) = \int_{-\infty}^{\infty} V \bar{S}_n(\vec{r}_0, V) dV, \tag{2.64}$$

where

$$\bar{S}_n(\vec{r}_0, V) = \bar{S}(\vec{r}_0, V) \bigg/ \int_{-\infty}^{\infty} \bar{S}(\vec{r}_0, V) dV \tag{2.65}$$

– the rated average density of a spectrum of power. The integral is equal in a denominator (2.65) to the total power of the reflected signal and can be received by integration (2.66) on volume

$$\int_{-\infty}^{\infty} \bar{S}(\vec{r}_0, V)dV = \bar{P}(\vec{r}_0) = \int_{V_i} \sigma(\vec{r}_n)I(\vec{r}_0, \vec{r}_n)dV_i. \quad (2.66)$$

The connection between velocities in the points $V(\vec{r}_n)$ and the moment weighed on power $\bar{V}_M(\vec{r}_0)$ (2.70) can be found by substitution of (2.63), (2.65) and (2.66) in the expression (2.64)

$$\bar{V}_M(\vec{r}_0) = \int_{V_i} V(\vec{r}_n)\sigma(\vec{r}_n)I(\vec{r}_0, \vec{r}_n)dV_i / \int_{Vi} \sigma(\vec{r}_n)I(\vec{r}_0, \vec{r}_n)dV_i. \quad (2.67)$$

The value (2.67) determined by a joint contribution of power (specific EER) and the function of radiation (2.61) can differ considerably from an average on space value of velocity of reflectors.

Similarly, we receive the expression for the weighed value of mean square width ΔV_M of a spectrum of radial velocities of HM

$$\Delta V_M^2(\vec{r}_0) = \int_{-\infty}^{\infty} \left[V - \bar{V}(\vec{r}_0)\right]^2 \bar{S}_n(\vec{r}_0, V)dV. \quad (2.68)$$

The expression (2.68) can be brought to the following form

$$\Delta V_M^2(\vec{r}_0) = \frac{\int_{V_i} V^2(\vec{r}_n)\sigma(\vec{r}_n)I(\vec{r}_0, \vec{r}_n)dV_i}{\int_{Vi} \sigma(\vec{r}_n)I(\vec{r}_0, \vec{r}_n)dV_i} - \left(\bar{V}_M(\vec{r}_0)\right)^2. \quad (2.69)$$

This dependence characterizes the weighed deviation of velocities from the average velocity.

The dependence of values $\bar{V}_M$ and ΔV_M on the form and the sizes of the allowed volume of the RS set by the radiation function and also on distribution of the specific EER in the allowed volume generally significantly complicates the problem of assessment of true values $\bar{V}$ and ΔV parameters of a spectrum of HM velocities on the basis of the analysis of the moments of DS of the reflected R signals.

2.4.3.2 Influence of the Allowed Volume Sizes on the Spectral Characteristics Assessment of the Radio Signal Reflected by MO

Let us assume that reflectivity $\sigma(\vec{r}_n)$ is a constant, and radiation function $I(\vec{r}_0, \vec{r}_n)$ is fixed and depends only on the difference $\vec{r}_0 - \vec{r}_n$, then (2.64) is transformed into the form

$$\bar{V}(\vec{r}_0) = \int\limits_{V_i} V(\vec{r}_n) I_n(\vec{r}_0 - \vec{r}_n) dV_i, \tag{2.70}$$

where $I_n(\vec{r}_0 - \vec{r}_n) = \frac{I(\vec{r}_0 - \vec{r}_n)}{\int\limits_{V_i} I(\vec{r}_0 - \vec{r}_n) dV_i}$ is the rated function of radiation.

Let us assume that the value of the allowed volume is small in comparison with the range so the divergence of radial velocities in the allowed volume can be neglected. Then, as the Eq. (2.70) represents convolution integral, the expression for a spatial spectrum $F_{\bar{V}}(\mathbf{K})$ of average radial velocities can be written in the form

$$F_{\bar{V}}(\mathbf{K}) = (2\pi)^6 F_X(\mathbf{K}) |F(\mathbf{K})|^2, \tag{2.71}$$

where $\mathbf{K}$—the spatial wave number;
$F_X(\mathbf{K})$—the spectrum (2.5) of radial velocities in the point;
$F(\mathbf{K})$—the Fourier transformation of the radiation function $I_n(\vec{\mathbf{r}}_0 - \vec{\mathbf{r}}_1)$.

Thus, at R measurement the spatial spectrum of turbulence is filtered by weight function of the allowed volume that leads to reduction of observed intensity of turbulence at scales, smaller than the radial size of the allowed volume.

For the majority of functions of radiation $I_n(\vec{r}_0 - \vec{r}_n)$ the good approximation of value $F(\mathbf{K})$ is three-dimensional Gaussian function [14, page 383]

$$|F(\mathbf{K})|^2 = (2\pi)^{-6} \exp\left[-K_X^2 \sigma_r^2 - \left(K_Y^2 + K_Z^2\right) r^2 \sigma_\alpha^2\right], \tag{2.72}$$

where K_X, K_Y and K_Z—the spatial frequencies of a turbulent flow along the axes OX, OY and OZ respectively;
σ_α^2—the second central moment of the ADP (for the radiation and reception) which for the ADP of a Gaussian type with circular symmetry can be presented by the expression

$$\sigma_\alpha^2 = \sigma_\beta^2 = \Delta\alpha^2/(16\ln 2) = \Delta\beta^2/(16\ln 2);$$

σ_r^2—the second central moment of function of spatial resolution which for pulse PS with the rectangular envelope and Gaussian TF of the receiver is equal to $\sigma_r^2 = (0{,}35 c\tau_{\text{и}}/2)^2$.

Having substituted the expression (2.71) in the ratios (2.6) and (2.7), we receive the equations for the measured radar stations (filtered by the allowed volume) of one-dimensional spectra of turbulence [14, page 383]

$$S_X(K_X) = \frac{(2\pi)^6}{2} \int\limits_{K_X}^{\infty} \left(1 - \frac{K_X^2}{K^2}\right) \frac{E(K)}{K} |F(K)|^2 dK, \tag{2.73}$$

$$S_Y(K_Y) = \frac{(2\pi)^6}{4} \int_{K_Y}^{\infty} \left(1 + \frac{K_Y^2}{K^2}\right) \frac{E(K)}{K} |F(K)|^2 dK, \tag{2.74}$$

$$S_Z(K_Z) = \frac{(2\pi)^6}{4} \int_{K_Z}^{\infty} \left(1 + \frac{K_Z^2}{K^2}\right) \frac{E(K)}{K} |F(K)|^2 dK. \tag{2.75}$$

Let us set the dispersion of turbulent velocity in the point through ΔV_p^2

$$\Delta V_p^2 = \langle v^2 \rangle - \langle v \rangle^2,$$

where angular brackets mean ensemble averaging.

At uniform structure of RR the width of a spectrum of radial velocity ΔV is defined as

$$\Delta V^2 = \overline{v^2} - (\bar{v})^2, \tag{2.76}$$

where the above line means spatial averaging of velocities taking into account the influence of DP and the sizes of the allowed volume.

The dispersion of average Doppler velocity at ensemble averaging has the form

$$\sigma_{\bar{v}}^2 = \langle (\bar{v})^2 \rangle - \langle \bar{v} \rangle^2. \tag{2.77}$$

Assuming the local uniformity of turbulence, axial symmetry of radiation function $I_n(\vec{\mathbf{r}}_0 - \vec{\mathbf{r}}_1)$ and considering that the sequence of performance of operations of ensemble averaging and space averaging can be changed, we will write down the Eq. (2.77) in the form

$$\sigma_{\bar{v}}^2 = \langle (\bar{v})^2 \rangle - \overline{\langle v \rangle^2}. \tag{2.78}$$

Then the ratio (2.76) taking into account (2.78) is brought to the form

$$\Delta V_p^2 = \langle \sigma_v^2 \rangle + \sigma_{\bar{v}}^2. \tag{2.79}$$

The expression (2.79) shows that the dispersion in the point is equal to the sum of spectrum width square at ensemble averaging and the spatial dispersion of velocities weighed by radiation function (the allowed volume).

Let us express the values of dispersion of the velocities measured in the point and average on the allowed volume through the corresponding characteristics of spectra

$$\Delta V_p^2 = \int F_X(\mathbf{K}) dV \quad \text{and} \quad \sigma_{\bar{v}}^2 = \int F_{\bar{v}}(\mathbf{K}) dV.$$

In this case the DS width taking into account (2.68) is equal to

$$\Delta V_p^2 = \int F_X(\mathbf{K})dV - \int F_{\bar{v}}(\mathbf{K})dV = \int \left[1 - (2\pi)^6 |F(\mathbf{K})|^2\right] F_X(\mathbf{K})dV. \tag{2.80}$$

Having substituted in (2.80) the expressions (2.5) and (2.72), we receive the following dependences

$$\Delta V_p^2 = \int_{V_и} \left[1 - \exp\left[-K_X^2\sigma_r^2 - \left(K^2 - K_X^2\right)r^2\sigma_\alpha^2\right]\right]\left(1 - \frac{K_X^2}{K^2}\right)\frac{E(K)}{4\pi K^2}dV \approx$$
$$\approx \left[1 - \exp\left[-K_X^2\sigma_r^2 - \left(K^2 - K_X^2\right)r^2\sigma_\alpha^2\right]\right]\left(1 - \frac{K_X^2}{K^2}\right)\frac{E(K)}{4\pi K^2}V_и.$$

At $E(K) = A\varepsilon^{2/3}K^{-5/3}$ the width of a spectrum of radial velocity will be equal to

$$\Delta V_p^2(K_X) = \left[1 - \exp\left[-K_X^2\sigma_r^2 - \left(K^2 - K_X^2\right)r^2\sigma_\alpha^2\right]\right]\left(1 - \frac{K_X^2}{K^2}\right)\frac{A\varepsilon^{2/3}K^{-11/3}}{4\pi}V_и.$$

As the vertical scale of turbulence is one order less than horizontal one [14], the contrary is true for spatial frequencies. Then $K^2 = K_X^2 + K_Y^2 + K_Z^2 \approx 2K_X^2$. In this case

$$\Delta V_p^2(K_X) = \left[1 - \exp\left[-K_X^2\sigma_r^2 - K_X^2 r^2\sigma_\alpha^2\right]\right]\frac{A\varepsilon^{2/3}K^{-11/3}}{8\pi}V_и. \tag{2.81}$$

In the publications [14, 44, 45] devoted to the analysis of the experimental data obtained with use of planes laboratories it is noted that the exponential summand in (2.81) in most cases, especially for airborne RS with high resolution, is almost insignificant therefore it is possible to consider the values of the corresponding weight coefficients in (2.70) and (2.82) as constants within the V_i volume. In this case the assessments of $\bar{V}_M$ and ΔV_M will also be a good approach to true values of $\bar{V}$ and ΔV. Therefore further we assume that

$$\bar{V} \approx \bar{V}_M, \qquad \Delta V \approx \Delta V_M. \tag{2.82}$$

2.4.3.3 Influence of the Probing Signal Repetition Period on the Assessment of the Spectral Characteristics of the Radio Signal Reflected by MO

The simultaneous independent influence of uniform WS and a set of the factors increasing the width of a spectrum of radial velocities (turbulence, movement of a beam, movement of the RS carrier, etc.) lead to the fact that the spectrum of power of the reflected signals can be believed to be Gaussian [14, page 124]

$$S(f) = \frac{S}{\sqrt{2\pi}\Delta f}\exp\left[-\frac{(f-\bar{f})^2}{2\Delta f^2}\right] + \frac{2NT_n}{\lambda}. \tag{2.83}$$

The corresponding ACF will have the form

$$B(mT_n) = S\exp\left[-2(\pi\Delta f mT_n)^2\right]\exp\left[-j2\pi\bar{f}mT_n\right] + N\delta m = S\rho(mT_n)\exp\left[-j2\pi\bar{f}mT_n\right] + N\delta m, \tag{2.84}$$

where

$\rho(mT_n) = \exp\left[-2(\pi\Delta f mT_n)^2\right]$—the rated correlation coefficient;
N—the spectral density of power of noise of the RS receiver.

The correlation between readings is significantly necessary for ensuring high accuracy of assessment of the DS parameters of the signal reflected by MO. The correlation coefficient (2.87) will increase at reduction of the period of repetition T_n. At the same time the dispersions of assessments of average frequency and RMS width of DS [14]

$$\mathrm{var}(\bar{f}) \approx \lambda^2\frac{(1+N/S)^2 - \rho^2(T_n)}{8M\rho^2(T_n)T_n^2} \quad \text{and}$$

$$\mathrm{var}(\Delta f) \approx \lambda^2\frac{\left(1-\rho^2(T_n)\right)^2 + 2\left(1-\rho^2(T_n)\right)N/S + \left(1+\rho^2(T_n)\right)(N/S)^2}{32\pi^2 M\Delta f\rho^2(T_n)T_n^2}$$

will decrease under the exponential law.

Let us define the limits of possible change T_n. According to [14] high correlation of readings is provided at correctness of inequality

$$\lambda/(2T_n) \geq 2\pi\Delta V, \tag{2.85}$$

i.e. the width of the spectrum of velocities of turbulent MO should be significantly less than the interval of unambiguous determination of velocity.

On the other hand, the reduction of the repetition period is limited by the need for unambiguous measurement of MO range within the RS coverage. The maximum value of unambiguously measured range is

$$r_{max} = cT_n/2. \tag{2.86}$$

Then from (2.85) and (2.86) the inequality defining the limits of permissible values of the period of repetition follows

$$2r_{max}/c \leq T_n \leq \lambda/(4\pi \Delta V). \tag{2.87}$$

2.4.4 *Influence of the Carrier Movement on Spectral and Power Characteristics of the R Signals Reflected by Meteoobjects*

2.4.4.1 Influence of the Carrier Movement on the Assessment of Average Frequency and Width of the Spectrum of MO Signals Received by an Airborne RS

The own movement of the RS carrier as it was noted in the Sect. 1.3 leads to additional movement of the disseminating MO particles in relation to ASFC and, therefore, to emergence in DS of the reflected signals of the additional components distorting the valid spectrum of radial speeds of reflectors. This circumstance causes the need of consideration and compensation of influence of own movement of the RS carrier at assessment of the Doppler spectrum parameters of signals.

Let's consider the influence of the radial and tangential components the RS carrier velocity. For the assessment of the contribution of these components we will introduce some restrictions:

1. The value of movement of the RS carrier during processing of signals should be much less than the range at which the assessment of parameters of the reflected signals is made.
2. The value of movement of the RS carrier should not exceed the sizes of the allowed volume of a RS in the corresponding direction.

Existence of a radial component of the RS carrier velocity leads to the shear of the whole DS of the reflected signals (without distortion of a form) along an axis of frequencies by the value $\bar{f}_{dv}$ (1.23). Besides, the high radial velocity of the movement of the RS carrier (at small MO elevation angles) also leads to considerable change of MO reflecting volume during processing of signals [34]. The change of spatial position of the allowed volume at the movement of the RS carrier leads to the change of structure of the reflectors forming a signal from V_i, and, therefore, to the decorrelation of the reflected signals which is expressed in widening of a spectrum of fluctuations of signals [34]. The interval of correlation of these fluctuations is equal to the time necessary in order that the allowed volume was completely updated.

$$\tau_f = \frac{-\tau_C}{2} \cdot \frac{1}{W_r}.$$

There is also another reason for the DS expansion of the signals accepted by a mobile RS. The regular cross (in relation to the radiation direction) movement of reflectors caused, for example, by tangential movement of the RS carrier leads to it [46, 47]. Let us consider the case of tangential movement of the RS carrier in the azimuthal plane (the influence of movement in the elevation plane is similar). Let the RS carrier move with the velocity W and the maximum of DR makes an angle $\alpha_0 \neq 0$ with the direction of the movement.

Each HM entering the allowed MO volume moves in relation to a RS with the radial velocity depending on its azimuth α_n, determined by the expression

$$V_n = W \cos \alpha_n.$$

Therefore, the dispersion of radial velocities of the reflectors entering the allowed volume is defined by the expression

$$\partial V = W \sin \alpha \partial \alpha,$$

which at narrow DP, i.e. in case of change of α in small limits when approximately directly proportional dependence remains between the velocity V_n and the azimuth α_n

$$\partial V = W_\tau \Delta \alpha,$$

i.e. the dispersion of velocities is defined by the product of the velocity of cross movement of the carrier and the width of DR from where the range of Doppler frequencies or the width of the spectrum of the fluctuations caused by cross movement of the RS carrier in the azimuthal plane is equal to

$$\Delta f_\tau = \frac{2\partial V}{\lambda} = \frac{2W_\tau \Delta \alpha}{\lambda}. \tag{2.88}$$

For example, at the tangential velocity of the RS carrier $W_\tau = 30\,\text{m/s}$, the wavelength $\lambda = 3$ cm and the DP width $\Delta\alpha = 2.6°$ the width of the spectrum of fluctuations $\Delta f_\tau \approx 100$ Hz, that is even a small deviation of DP from the direction of the carrier ($W_\tau = 30\,\text{m/s}$ at W = 600 m/s and $\alpha_0 = 3°$) leads to noticeable widening of the spectrum of the reflected signal.

If there are tangential movements of the RS carrier at the same time in two planes (azimuthal and elevation), then the width of the spectrum of fluctuations (2.88) in this case will be equal to

$$\Delta f_\tau = \sqrt{\Delta f_y^2 + \Delta f_z^2} = \frac{2}{\lambda}\sqrt{\left(W_y \Delta\alpha\right)^2 + (W_z \Delta\beta)^2}, \tag{2.89}$$

where W_y, W_z are carrier velocity projections to the axes OY_a and OZ_a respectively.

Let us estimate the influence of movement of the RS carrier of type of the ACF signals reflected by MO and accepted by mobile RS. The presence of nonzero radial velocity of rapprochement of the RS carrier with the studied allowed volume leads to emergence in the reflected signals of the additional phase shift characterized by frequency (2.89). The tangential movement of the carrier causes reduction of the ACF module because of its dependence on DP values. In the assumption of separability of DP we approximate it by widely used Gaussian functions

$$G(\alpha_n) = \exp\left(-\frac{(\alpha_n - \alpha_0)^2}{\Delta\alpha}\right); \quad G(\beta_n) = \exp\left(-\frac{(\beta_n - \beta_0)^2}{\Delta\beta}\right). \tag{2.90}$$

Then, owing to moving of the RS carrier to ACF of the reflected signals (2.37), the following factor will appear

$$B_{ЉЂ‰}(\tau) = \exp\left(j\frac{4\pi\Delta_r}{\lambda}\right)\exp\left(-\frac{2\pi^2\Delta_y^2\Delta\alpha^2}{\lambda^2}\right)\exp\left(-\frac{2\pi^2\Delta_z^2\Delta\beta^2}{\lambda^2}\right), \tag{2.91}$$

where Δ_r—the movement of the RS carrier in the radial direction in time τ; Δ_y, Δ_z—the movements of the carrier for the time τ along the axes OY_a and OZ_a respectively.

For the exclusion of fluctuations of signals arising because of radial and tangential movement of the RS carrier, it is necessary to stabilize the position of the allowed volume in space in relation to the RS APC or, in other words, to displace APC so that its position in relation to the centre of the allowed MO volume during time of processing of the reflected signals would remain constant (the so-called "RS quasimotionless" mode).

Let's determine the APC shift value. For this purpose it is necessary to estimate the corresponding projections of velocity of change of the position of the allowed volume concerning APC due to the movement of the RS carrier. These projections are equal to:

- the radial projection $W_r = W \cos\alpha_0 \cos\beta_0$;
- the projection to the axe OY_a $W_y = W \sin\alpha_0$;
- the projection to the axe OZ_a $W_z = W \cos\alpha_0 \sin\beta_0$.

Then for the time τ the relative movement of the allowed volume will be:

- the radial projection $\Delta_r = W\tau \cos\alpha_0 \cos\beta_0$;
- along the axe OY_a

$$\Delta_y = W\tau \sin\alpha_0; \tag{2.92}$$

- along the axe OZ_a

$$\Delta_z = W\tau \cos\alpha_0 \sin\beta_0.$$

Taking into account (2.92) the expression (2.91) for a multiplier of ACF caused by the movement of the RS carrier will have the form

$$B_{JБ‰}(\tau) = \exp\left(j4\pi\frac{W\tau\cos\alpha_0\cos\beta_0}{\lambda}\right)\exp\left(-2\pi^2\left(\frac{W\tau\sin\alpha_0\Delta\alpha}{\lambda}\right)^2\right)\times$$
$$\times\exp\left(-2\pi^2\left(\frac{W\tau\cos\alpha_0\sin\beta_0\Delta\beta}{\lambda}\right)^2\right). \tag{2.93}$$

The system of coordinates $X_aY_aZ_a$ corresponds to the case of use of the AS with mechanical scanning in the horizontal and vertical planes. However, if the AS is rigidly connected with the axis X that takes place when using the phased array antennas (PAA) with electronic scanning, then it is necessary to pass to the system of coordinates XaYZ. For this case we receive

$$\Delta'_r = W\tau \sec\alpha_0 \sec\beta_0;$$
$$\Delta'_y = W\tau\, tg\,\alpha_0;$$
$$\Delta'_z = W\tau \sec\alpha_0 tg\,\beta_0. \tag{2.94}$$

Then the expression for a multiplier of ACF caused by the movement of the RS carrier with a PAA will have the form

$$B'_{JБ‰}(\tau) = \exp\left(j4\pi\frac{W\tau\sec\alpha_0\sec\beta_0}{\lambda}\right)\exp\left(-2\pi^2\left(\frac{W\tau\,\mathrm{tg}\,\alpha_0\Delta\alpha}{\lambda}\right)^2\right)\times$$
$$\times\exp\left(-2\pi^2\left(\frac{W\tau\sec\alpha_0\,\mathrm{tg}\,\beta_0\Delta\beta}{\lambda}\right)^2\right). \tag{2.95}$$

Let the following ratios be correct for RS AS

$$\Delta\alpha \approx \lambda/L_y, \qquad \Delta\beta \approx \lambda/L_z,$$

where L_y, L_z—the linear sizes of a RS AS aperture in the horizontal and vertical planes respectively.

If we introduce the following designations

$$\bar{f}_{\partial e} = \frac{2\Delta_r}{\lambda\tau}, \Delta f_y = \frac{\Delta_y \Delta\alpha}{\lambda\tau} \approx \frac{\Delta_y}{L_y\tau}, \Delta f_z = \frac{\Delta_z \Delta\beta}{\lambda\tau} \approx \frac{\Delta_z}{L_z\tau},$$

the expression (2.95) can be written down in the form

$$B_{dv}(\tau) = \exp(j2\pi \bar{f}_{dv}\tau)\exp(-\pi(\Delta f_y + \Delta f_z)\tau). \tag{2.96}$$

So, the ACF of the reflected signals accepted by radar station on the moving carrier and, respectively, their DS differ from the ACF and the spectrum of the signals accepted by motionless radar station. Besides the parameters $\bar{f}$ and Δf, depending on the parameters $\bar{V}$ and ΔV the spectrum of radial velocities of HM, the ACF of the signals accepted by mobile radar station will also depend on the parameters $\bar{f}_{dv}$, Δf_y, Δf_z, determined by own movement of the RS carrier. At the same time Δf_y and Δf_z cause the spreading of the DS of the reflected signals, and $\bar{f}_{dv}$ causes the shift of its average frequency. The ACF rated modules for different types of radiation and installation sites of RS are presented in Fig. 2.10 as an example.

As numerous experimental data [48, 49] testify,

$$\Delta f_y >> \Delta f, \quad \Delta f_z >> \Delta f.$$

These inequalities show that it is impossible to estimate with high precision the value of Δf and to find on it the presence of a zone of dangerous turbulence, with-

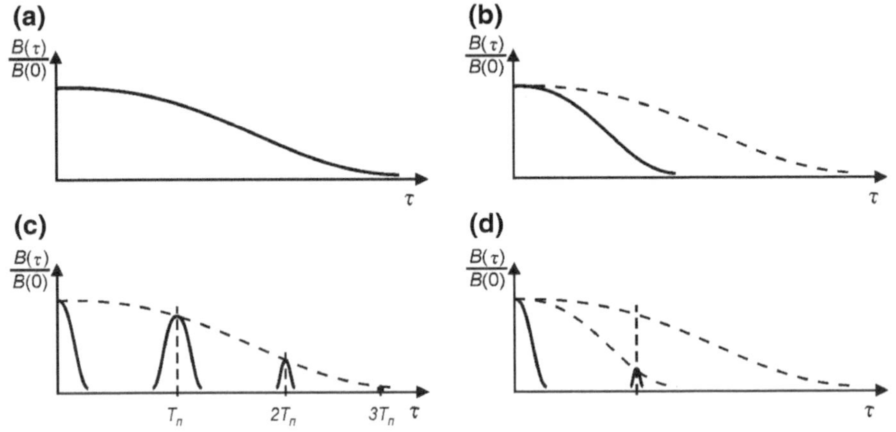

Fig. 2.10 The type of the rated ACF envelope of the signal reflected by the allowed MO volume: **a** continuous radiation and motionless RS; **b** the RS is located on the moving carrier; **c** pulse radiation and motionless RS; **d** the RS is located on the moving carrier

out having compensated the movement of APC caused by a tangential component of velocity of the RS carrier therefore it is necessary to make the conclusion that at detection of dangerous MO zones it is necessary to solve at the same time 2 interconnected problems:

- to make assessment of the moments of the spectrum of radial velocities of MO particles;
- to carry out compensation of influence of movements of the RS carrier.
- From (2.96) it follows that for exclusion of influence of own movement of the RS carrier it is necessary to provide the fulfilment of the condition

$$\bar{f}_{dv} = 0, \quad \Delta f_y = 0, \quad \Delta f_z = 0. \tag{2.97}$$

2.4.4.2 Influence of the Carrier Movement on the Assessment of Radar Reflectivity of MO

As it was noted in the Sect. 2.4.4.1, generally because of correlation of the reflected signals the mathematical expectation of the fluctuating part of power of the reflected signals (2.58) is not equal to zero that increases the RR assessment error. One of the factors defining the degree of correlation of signals at placement of RS onboard the mobile carrier are the velocity and the direction of the movement of the RS carrier.

In particular, taking into account (2.88) the coefficient of correlation (rated ACF) of the accepted signals can be expressed in the form of

$$\begin{aligned}\rho(\tau) &= \exp\left\{-\left[(2\pi\Delta f\tau)^2 + (2\pi\Delta f_\tau \tau)^2\right]\right\}\\ &= \exp\left\{-\left[\left(\frac{4\pi\Delta V\tau}{\lambda}\right)^2 + 4\pi\left(\frac{W\tau\Delta\alpha\sin\alpha_0\cos\beta_0}{\lambda}\right)^2 + 4\pi\left(\frac{W\tau\Delta\beta\sin\beta_0}{\lambda}\right)^2\right]\right\}.\end{aligned} \tag{2.98}$$

At accumulation of M readings of the reflected signals taken with the interval T_p, the correlation degree between them will be different owing to increase in a delay between them. For the m-th reading ($m \le M$) the delay time will be mT_p, and the expression (2.102) is transformed to the form

$$\rho(mT_n) = \exp\left[-m^2\left\{\left(\frac{4\pi\Delta VT_n}{\lambda}\right)^2 + 4\pi\left(\frac{WT_n\Delta\alpha\sin\alpha_0\cos\beta_0}{\lambda}\right)^2 + 4\pi\left(\frac{WT_n\Delta\beta\sin\beta_0}{\lambda}\right)^2\right\}\right],$$

whence it follows

$$\rho(mT_n) = [\rho(T_n)]^{m^2}. \tag{2.99}$$

Knowing the values of the correlation coefficient $\rho(mT_n)$, on (2.60) it is possible to define the number of effective (uncorrelated) readings M_E. The M_E values in the

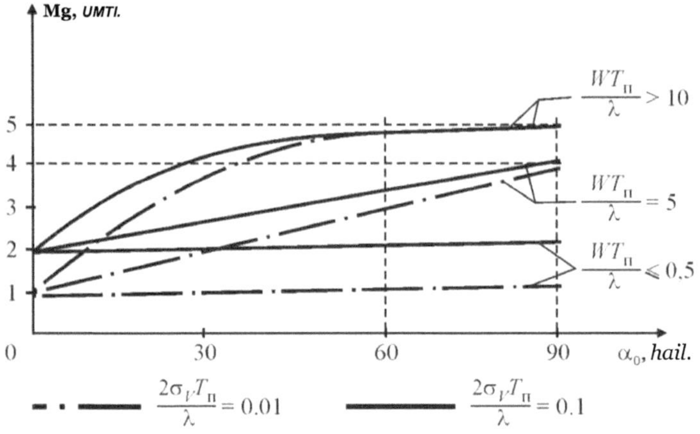

Fig. 2.11 Dependence of the size of effective selection of signals from the azimuth (for M = 5, $\Delta\alpha = \Delta\beta = 2.4$ deg.)

azimuth function α are presented in the Fig. 2.11 (at the same time it was assumed that $M = 5$, $\beta = 0$). The analysis of the dependences given in Fig. 2.11 shows that at $\frac{WT_n}{\lambda} > 10$ on the viewing angles α > 30° the full decorrelation of readings of the accepted signals takes place and $M_E = M$. Knowing the M_E values, it is possible to define the RMSD (error) of the assessment of average power $\bar{P}_{np}$ depending on an azimuth α.

At coherent accumulation of ME of the consecutive partially correlated readings of the reflected signals the RMS error of the assessment of average power (2.62) is equal to

$$\sigma_{\overline{P}} = \frac{\sigma_{Pm}}{M_E}, \tag{2.100}$$

where $\sigma_{\overline{P}}$—an RMS error of the assessment of average power;
σ_{Pm}—an RMS error of single measurement $P_{np\ m}$, which, as it is shown in the work [50, page 91], is 5.57 dB.

The values of RMSD $\sigma_{\overline{P}}$, calculated by the formula (2.100) for the M_E values presented in Fig. 2.11 are given in Fig. 2.12. The analysis of the dependences given in Fig. 2.12 illustrating the influence of own movement of the RS carrier on power characteristics of the signals reflected from MO showed that the error of the assessment of average power of the reflected signal and, therefore, RR due to the influence of the movement and correlation of selections lies within 2–2.2 dB at $WT_n/\lambda > 10$ and the azimuth $\alpha_0 > 30°$ (M = 5), and 3–5.6 dB at $WT_n/\lambda < 10$ and the azimuth $\alpha_0 < 30°$ (M = 5). Varying the value of the accepted pack M (due to change of the period of repetition of PS or the velocity of scanning of DP), it is possible to lower the RR assessment errors to the required value.

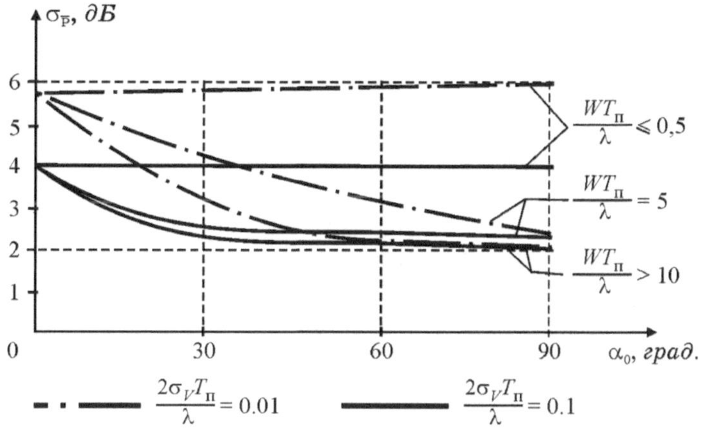

Fig. 2.12 Dependence of the RMS error of the assessment of average power on the azimuth (for M = 5, $\Delta\alpha = \Delta\beta = 2.4$ deg.)

Then it is necessary to determine the size of an emitted pack of impulses M for obtaining the constant set size of an effective pack M_E. The constancy of ME value on all azimuths in a front hemisphere of an AV ($\alpha_0 \leq \pm 90°$) will provide the constancy of an RMS error $\sigma_{\overline{P}}$ of assessment of RR on all azimuths. Let's transform the expression (2.60) taking into account (2.99)

$$\frac{1}{M_E} = \frac{1}{M} + \frac{2}{M}\sum_{m=1}^{M-1}\left(1 - \frac{m}{M}\right)\rho(T_n)^{m^2}. \tag{2.101}$$

For determination of size of an emitted pack M it is necessary to solve the Eq. (2.101) in relation to M at a preset value of an effective pack M_E. Owing to complexity of the analytical solution of this equation the assessment of M is made by numerical methods taking into account the following technical characteristics of a radar station:

- the wavelength $\lambda = 3.2$ cm;
- the frequency of repetition of PS $F_p = 1/T_p = 1$ kHz;
- the effective number of the accumulated impulses $M_E = 5$.

The result of calculation is presented in Fig. 2.13. Figure 2.13, a corresponds to a turbulence case with the RMS width of the spectrum of velocities $\sigma_V = 0.25$ m/s, Fig. 2.13, b—the turbulence with $\sigma_V = 2.5$ m/s.

(a) in case of turbulence with the RMS width of the spectrum of velocities $\Delta V =$ 0.25 m/s;
(b) in case of turbulence with the RMS width of the spectrum of velocities $\Delta V =$ 2.5 m/s

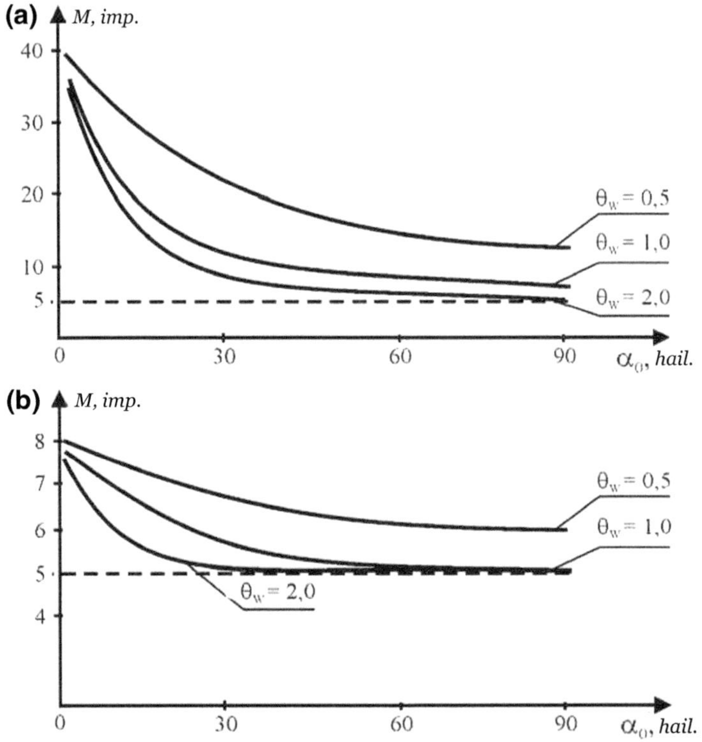

Fig. 2.13 The dependence of the size of the processed pack of signals for receiving of effective selection of ME

The value $\theta_W = 2\sqrt{2\pi}WT_n\Delta\alpha / \lambda$, is used as the parameter at calculation, at the same time:

- $\theta_W = 0.5$ corresponds to the combination $\Delta\alpha = 3°$, W = 450 km/h or $\Delta\alpha = 7°$, W = 200 km/h;
- $\theta_W = 1$ corresponds to the combination $\Delta\alpha = 3°$, W = 900 km/h or $\Delta\alpha = 7°$, W = 400 km/h;
- $\theta_W = 2$ corresponds to the combination $\Delta\alpha = 7°$, W = 800 km/h.

The analysis of the results of calculation given in Fig. 2.13 shows that the value of an emitted pack of impulses significantly depends practically on all characteristics of MO, the RS carrier and the RS, including on the viewing angles in the horizontal and vertical planes (from the azimuth and elevation angle).

2.4.5 The Use of Parametric Models for the Description of the Signal Reflected by MO

The time of R contact with MO of airborne RS of modern and perspective high-speed AV is extremely limited that does not allow to receive the reflected signal (a pack of the radio impulses reflected by MO) of large volume. The small duration of a pack (selection) owing to a fundamental ratio of uncertainties (the Sect. 2.5) determines the insufficient accuracy of measurement and RS resolution by the frequency (velocity).

The pack nature of the reflected signal is equivalent to its multiplication in a time domain by a rectangular window. At the same time it is supposed that a signal and its ACF are equal to zero outside an observation interval. Owing to such multiplication due to Gibbs effect the side lobes with the maximum level near −13 dB appear in the spectrum of the processed signal [33]. For smoothing of transition to zero values the discrete readings of a complex signal envelope (or its ACF) are exposed to weight processing before the Doppler analysis [51]

$$\dot{S}_w(m) = W(m)\dot{S}(m),$$

where W(t)—the function of a weight window.

Application of the weight processing directed to reduction of the masking action of side lobes of the spectrum leads to widening of the main lobe of the Doppler spectrum, i.e. to overestimation of the width of the spectrum of YM velocities and also to power losses and need of allocation of temporal resources on weight processing.

In practice, in particular, when signals are narrow-band and their ACF it is close to periodic, the specified assumption is not true. Therefore it is reasonable, having a small piece of ACF (or the small volume of selection of the reflected signal), to continue it out of observation interval limits, based on some prior data on the nature of the accepted signal [52]. In many practical cases the target situation causing formation of the reflected signal can be rather precisely simulated by the linear system of a finite order (the Annex 3). Generally the connection between input and output signals in a linear system (the forming filter) with the rational transfer function is described by the linear differential equation

$$s(m) = \sum_{k=1}^{p} a_k s(m-k) + \sum_{k=1}^{q} b_k e(m-k) + e(m), \tag{2.102}$$

where $s(m)$, $e(m)$—the output and entrance system signals respectively;
p—the order of autoregression (AR);
q—the order of moving average (MA);
a_k and b_k—the complex AR and MA coefficients respectively.

The model (2.102) is known as the model of autoregression—moving average (ARMA).

If in the expression (2.102) all coefficients $b_k = 0$, then the signal s(m) represents a linear regression of its previous values (the autoregression model)

$$s(m) = \sum_{k=1}^{p} a_k s(m-k) + e(m) \tag{2.103}$$

or in a matrix form

$$\vec{a}^T \vec{S}(m) = \vec{b}^T \vec{E}(m), \tag{2.104}$$

where T—the transposing operation symbol;
$\vec{a} = \left[1\ -a_1\ \ldots\ -a_p\right]^T$—(p + 1)—the element vector of AR coefficients;
$\vec{S}(m) = \left[s(m)\ s(m-1)\ \ldots\ s(m-p)\right]^T$—(p + 1)—the element vector of readings of an output signal;
$\vec{b} = \left[1\ b_1\ \ldots\ b_q\right]^T$—(q + 1)—the element vector of the MA coefficients; $\vec{E}(m) = \left[e(m)\ e(m-1)\ \ldots\ e(m-q)\right]^T$—(q + 1)—the element vector of readings of an input signal.

Respectively, if in the expression (2.102) all coefficients ak = 0, then this is the model of moving average.

The AR model in many practical cases can describe rather precisely the target situation causing formation of the reflected signal. It is connected with its ability to display adequately the narrow-band spectral components. Other models on the basis of rational functions (MA and ARMA) have no specified advantage or are now insufficiently investigated.

In case of use as the forming signal e(m) of discrete complex normal white noise with zero average value and dispersion σ_e^2 the expression connecting the parameters of the AR model with the ACF of the signal s(m) reflected by MO has the form [53, 54]

$$B(m) = \begin{cases} B^*(-m), & m < 0; \\ \sum_{k=1}^{p} a_k B^*(k) + \sigma_e^2, & m = 0; \\ \sum_{k=1}^{p} a_k B(m-k), & m > 0. \end{cases} \tag{2.105}$$

It follows from the expression (2.105) that when using the AR of model there is an opportunity to continue unambiguously the ACF of the signal s(t) indefinitely by means of the recurrence ratio

$$B(m) = \sum_{k=1}^{p} a_k B(m-k) \text{ for all m} > \text{q}. \tag{2.106}$$

The ACF Fourier transformation (2.105) represents the spectral density of power of the signal of s(m)

$$S(f) = \sigma_e^2 T_n \left| 1 - \sum_{k=1}^{p} a_k \exp(-j2\pi f k T_n) \right|^{-2} \tag{2.107}$$

or in a vector form

$$S(f) = \frac{\sigma_e^2 T_n}{\vec{c}_p^H(f)\vec{a}\vec{a}^H\vec{c}_p(f)}, \tag{2.108}$$

where $\vec{c}_p(f) = \left[1\ \exp(j2\pi f T_n) \ldots \exp(j2\pi f p T_n)\right]^T$—the vector of complex sinusoids;

N—the symbol of the operation of Hermite conjugation consisting consecutive performance of operations of transposing and complex conjugation.

It follows from the expression (2.107) that for the assessment of S(f) by processing of the selection of real signals [52] it is necessary to define the values a_k of AR coefficients and also the dispersion σ_e^2 of the forming signal (noise).

If the polynom which is in a denominator of the spectral density (2.107) is factorized, then it can be written down in the form

$$S(f) = \sigma_e^2 T_n \left(\prod_{k=1}^{p} |A_k(f)| \right)^{-2}, \tag{2.109}$$

where $|A_k(f)| = \sqrt{\mathrm{Re}\{A_k(f)\}^2 + \mathrm{Im}\{A_k(f)\}^2}$;
$A_k(f) = 1 - |z_k| \exp[-j(2\pi f T_n - F_k)]$—the spectrum of the AR process of the first order;
$z_k = |z_k| \exp(jF_k)$—the poles of the AR model (Annex 3).

As the signals reflected by MO can be adequately presented by an AR process of low (second or fifth) order [55–57], the definition of poles of the corresponding AR model can be executed by one of effective methods of calculation of roots of polynoms (the method of the analysis of own values of the companion matrix, Laguerre, Lobachevsky methods, etc.). One of the effective methods of search of roots of a characteristic polynom is the search of own numbers of the corresponding matrix. This statement formed the basis of thee method of assessment of complex poles of the model on the known values of AR coefficients by search of own values of the companion matrix [58]

$$\underline{\mathbf{A}}_{comp} = \begin{bmatrix} -\frac{a_{p-1}}{a_p} & -\frac{a_{p-2}}{a_p} & \dots & -\frac{a_1}{a_p} & -\frac{a_0}{a_p} \\ 1 & 0 & \dots & 0 & 0 \\ 0 & 1 & & 0 & 0 \\ \dots & \dots & & \dots & \dots \\ 0 & 0 & \dots & 1 & 0 \end{bmatrix}. \quad (2.110)$$

One of the rapidly converging numerical algorithms of search of own values of the complex matrix is the QR algorithm with implicit shift [58].

If we use the AR process of the first order as a model of the MO signal, then its spectrum (2.107) symmetric in relation to $\bar{f}$, is defined by the expression (Annex 3)

$$S(f) = \frac{\sigma_e^2 T_n}{|1 - a_1 \exp(-j2\pi f T_n)|^2} = \frac{\sigma_e^2 T_n}{1 + |a_1|^2 - 2|a_1| \cos(2\pi f T_n - \Phi)},$$

where $a_1 = |a_1| \exp(jF)$—the unknown complex coefficient of the AR model.

If the AR coefficient module $|a_1| \to 1$, then the signal reflected by MO is the narrow-band random sequence which spectrum average frequency coincides with the spectrum maximum frequency [59]

$$\bar{f} = F/(2\pi T_n) = Arg(a_1)/(2\pi T_n), \quad (2.111)$$

and the expression for the spectrum RMS width after a number of transformations is brought to the form [60]

$$\Delta f = \frac{1}{2\pi T_n} \left[\frac{\pi^2}{3} - 4 \sum_{k=1}^{\infty} \frac{(-1)^{k+1}}{k^2} |a_1|^k \right]^{1/2}. \quad (2.112)$$

Thus, the assessment of AR coefficient a_1 is enough for measurement of $\bar{f}$ and Δf.

In case of approximation of the spectrum by AR models of any order the received expressions have very simple physical interpretation: each AR coefficient corresponds to existence in the reflected signal of the narrow-band random sequence with its average frequency and spectrum width. This takes place in case MO consists of several groups of the reflectors moving with various velocities [61, 62]. The selection of separate spectral components is, as a rule, made by definition of local maxima of multimode DS of the accepted signals. Further on the basis of the assessment of intensity of the corresponding modes of the spectrum, the decision is made on which of the AR model poles should be considered.

Here it should be noted that the important parameter of the ARMA, AR and MA models is the model order. Provision of a compromise between resolution and accuracy (dispersion) of the received spectral assessments depends on the choice of the order [54, page 255]. As it is noted in [55–57], the signals reflected by MO can be adequately presented by a random process of a low (second or fifth) order. If the order of model is chosen as too small, then strongly smoothed spectral estimates are received. For decrease in dispersion of estimates in this case it is necessary to accumulate big selections of signals (about 1000) that is unacceptable in case of use of an airborne RS. The overdetermination of the model (excessive overestimation of an order) in the conditions of existence of errors of measurement can lead to additional, often very essential, errors in the assessment of the spectrum [54, 63]. In particular, emergence of false maxima in the spectrum is possible.

Several various criteria (criterion functions) (the Annex 3) are offered for the choice of an order of the model, however the results of estimation of spectra when using these criteria does not differ significantly from each other, especially in case of processing of real signals, but not modelled processes with the set statistical properties [54, page 283]. And, as it is proved in [64, 65], in case of processing of selections of signals of limited volume any of the given criteria does not provide satisfactory results therefore in the analysis of short packs of the reflected signals it is the most expedient to choose the order of the AR model in the range $M/8 \le p \le M/2$, that is experimentally confirmed in [66].

Determination of the AR model parameters of the signal reflected by the allowed MO volume is considerably complicated by two factors. Firstly, the significant growth of uncertainty in the choice of the model order due to distinction of Doppler shifts of frequency of the reflected signals accepted in the main and side lobes of the ADN, dependence of width of the spectrum of reflections in the main lobe from the sight direction and presence of local maxima of the spectrum in the field of side lobes that, in principle, leads to substantial increase of the model order with the corresponding increase in computing expenses at calculation of weight coefficients of the forming filter. Secondly, the existence of connection between spatial and Doppler frequencies of the signal for the airborne RS which is absent at motionless radars [34].

Summing up the result, it should be noted that:

1. The signal reflected by any allowed MO volume represents the superposition of mutually independent partial signals of separate HM. The distribution of the amplitude of the total signal is connected with distribution of the specific EER on the MO volume and is described by Rayleigh law and the distribution of a random initial phase is uniform in the interval $[-\pi,\pi]$.
   The specified signal after linear analog processing and quadrature phase detecting in the RS receiver is transformed in the ADC block to discrete readings of quadrature components of the complex envelope, representing two statistically connected discrete random processes. If the radial component of average wind speed $\bar{V} = 0$, then the specified processes can be considered as mutually independent Gaussian random processes with zero mathematical expectation and the dispersion equal to average power of the reflected signal. In this case the mathematical model of the signal reflected by the MO allowed volume represents the model of two independent Markov processes of the first order.
   The nonzero radial component of average wind speed leads to emergence of correlation between these random processes, and the module of coefficient of mutual correlation depends on the value $\bar{V}$, and the sign—on the sign (direction)

of speed. At the same time the quadrature components of the complex envelope the reflected signal become the correlated Markov processes of the second order.

2. The dependence of assessments of values of average velocity and width of the spectrum of velocities of reflectors on the form and values of the RS allowed volume and also on distribution of the specific EER in the allowed volume generally significantly complicates the problem of determination of true values of the corresponding parameters. However, in the majority of practical cases, especially for airborne RS with high resolution, it is possible to consider the fields of radial velocity and specific EER as locally stationary within the V_i volume.
3. The influence of AV own movement at assessment of RR affects only the quantity of M_E of uncorrelated selections of a signal. At the same time the value of M_E significantly depends on the sight angles α_0, β_0 and the ratio of the distance passed by the RS carrier during repetition to a wavelength.
4. The essential limitation of R contact of an airborne RS with MO does not allow to receive the reflected signal (pack) of large volume and by that to provide the required accuracy of measurement and resolution of a radar station on the Doppler frequency (velocity). The packed nature of the reflected signal leads to emergence in its spectrum at the expense of Gibbs phenomenon of side lobes with the maximum level near −13 dB. The application of the weight processing directed to reduction of the masking action of side lobes of the spectrum causes the widening of the DS main lobe, i.e. leads to overestimation of the width of the spectrum of YM velocities and also to power losses and need of allocation of time resources on weight processing.
5. In many practical cases the target situation causing formation of the reflected signal can be rather precisely simulated by the linear system of a finite order (in particular, the autoregression model). In this case for the assessment of the spectrum of the reflected signal it is necessary to define the values of AR coefficients and dispersion of the excitation signal. The spectra of signals approximated by AR models have very simple physical interpretation: each AR coefficient corresponds to existence in the reflected signal of the narrow-band random sequence with its average frequency and spectrum width. This takes place in case MO consists of several groups of the reflectors moving with various velocities.

The important parameter of AR models is the model order. Provision of a compromise between resolution and dispersion of the received spectral assessments depends on the choice of the order. The signals reflected by MO can be adequately presented by a random process of a low (second or fifth) order.

2.5 The Mathematical Model of a Path of Processing of Signals of an Airborne RS

The problem of reliable detection of dangerous MO zones in a front hemisphere at big and average range, measurement of their polar coordinates and also the assessment of degree of their danger is one of the main for airborne meteo RS of AV of civil aviation [67–71]. Though for airborne RS of modern and perspective AV of the state aircraft the specified task is not the main, it is very important for ensuring their round-the-clock and all-weather application. At the same time, considering the need of performance in the course of R observation of a large number of diverse operations, in the considered mode, by analogy with other modes of airborne RS, it is necessary to allocate the stages of primary and secondary information processing [72, 73]. The view of space, selection, permission, detection and assessment of parameters of the signals reflected by MO for making decision on existence of potentially dangerous zones of MO and definition of their position in the BCS belong to preprocessing (processing of signals). The assessment of the influence of the movement of the RS carrier and its compensation also belong to preprocessing. The secondary processing (data processing) is made on the basis of the results of processing of signals, information on the current location and movement of the carrier for the purpose of formation of assessments of meteorological parameters of the MO areas. At the same time meteorological information is taken from spatial distributions (fields) of the measured R values.

The essence of modelling of a path of preprocessing of signals of airborne RS by method of the complex envelope consists in replacement of a radio system with narrow-band input influences with the model—an equivalent low-frequency complex system with input influence in the form of the complex envelope of a narrow-band signal. The model should include the ratios allowing to calculate the complex envelopes of narrow-band signals and interference at the system output on instantaneous values of corresponding complex envelopes at input. In this case modelling is made according to the block diagram of standard airborne radar station (Fig. 2.14) which includes a synthesizer of frequencies, a transmitter, an antenna system (AS), a receiver, a ADC block, a reprogrammable signal processor (RSP) and a data processor (the onboard digital computer, OBC) connected among themselves by the information highway—the intermodular parallel interface (IPI) [58]. In case of use of the active phased antenna array (APAA) a transmitter, AS and a radio-frequency part of the receiver are structurally and functionally integrated [74, 75].

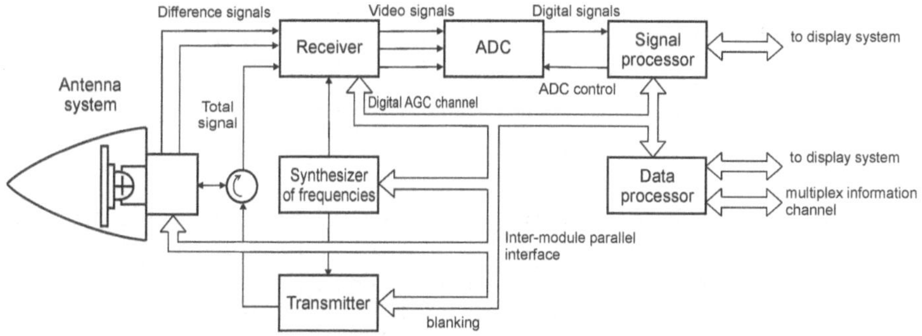

Fig. 2.14 Block diagram of a typical aircraft radar

The synthesizer of frequencies carries out formation of the probing and basic fluctuations for the transmitter and the receiver of a radar station. The PS amplified by the transmitter are radiated in the form of spherical EMW by the AS to open space. In modern airborne RS the greatest distribution was gained by the PS in the form of the coherent sequence of simple or complex radio impulses. The complex envelope of a coherent pack of the probing impulses has a form of convolution [76]

$$S_u(t) = \sqrt{P_u}\left[B_0(t) \ll \times \gg \sum_{m=0}^{M-1} \delta(t - mT_n)\right], \qquad (2.113)$$

where $B_0(t)$—the complex envelope of the single probing radio impulse;
$\ll \times \gg$—the symbol of convolution operation;
$\delta(t) = \begin{cases} \infty & when\ \mathrm{t} = 0, \\ 0 & when\ \mathrm{t} \neq 0. \end{cases}$—the delta function.

At the coordinated intra period processing the width of a spectrum of PS determines the accuracy of measurement and resolution of a radar station by the range (delay time) [77, 78]

$$\delta r = \frac{c\delta t_з}{2} \approx \frac{c}{4\Delta f_м}\sqrt{\frac{N}{2E}}, \qquad (2.114)$$

and the effective duration of PS—the accuracy of measurement and resolution of radar station on velocity (on Doppler frequency) [77, 78]

$$\delta f_д \approx \frac{1}{\tau_{эф}}\sqrt{\frac{N}{2E}}. \qquad (2.115)$$

Here $\Delta \mathrm{f}_м$—the width of a spectrum of modulation of the probing signal; N—the spectral density of noise; E—the energy of a signal.

The product of errors of independent measurements of range (delay time) (2.114) and velocity (Doppler frequency) (2.115)

$$\delta t_з \delta f_д \approx \frac{1}{2\Delta f_м \tau_{эф}}\frac{N}{2E}. \qquad (2.116)$$

At simple pulse signals duration and width of a spectrum are interconnected ($\tau_{ef}\Delta \mathrm{f}_м \approx 1$), therefore at invariable energy of a signal it is possible to increase the accuracy of measurement of velocity, only having reduced the accuracy of measurement of range. The spectrum of complex signals is defined by additional modulation of amplitude, frequency or phase of the carrier fluctuation and practically does not depend on duration therefore the effective duration and effective width of a spectrum of complex signals can be changed independently of each other. At the same time the base of the signal $\tau_{ef}\Delta f_{м}$ can be made an essentially big value.

According to the specified tactical technical requirements provided in the Sect. 1.4 to the airborne radar station functioning in the modes of detection and assessment of danger of WS and turbulence at SNR of 28 … 30 dB the accuracy of measurement of range of dangerous MO should be 50 … 100 m, the accuracy of measurement of velocity $\pm$ 1 m/s. The substitution of the specified values in (2.116) shows that the required base of a signal is about a unit that confirms the possibility of use in the considered PS modes in the form of the coherent sequence of simple radio impulses without intra pulse modulation.

So, it is necessary to consider the following PS parameters at modelling:

- the duration of the probing radio impulse τ_u;
- the quantity of impulses in the pack M;
- the period of repetition of impulses T_n;
- the impulse emitted power P_u;
- the wavelength λ.

EMW corresponding to impulse PS of a radar station are in free space. In a distant zone the emitted EMW can be considered to be approximately flat. In electromagnetic interaction of the falling (radiated) wave and the object which is in a view zone of a system the reflected EMW extending in the direction of a radar station is formed. It is characterized by two spatial parameters (the angular coordinates defining the direction of its arrival to radar station), two parameters describing its polarizing structure (for example, shift of phases of orthogonal components and the ratio of their amplitudes) and four parameters defining its change in time (frequency, amplitude, initial phase and beginning of time marking).

The characteristics of a RS as a uniform polarizing and space-time- filter should be agreed with the corresponding parameters of reflected EMW. We will further believe that polarizing and space-time processing of signals is separable, i.e. divisible on polarizing, spatial and temporal processing. In airborne radar stations with rather narrow-band PS and AS small apertures this condition is practically always satisfied [79].

The reflected EMW are accepted by the RS AS carrying out polarizing and spatial filtration of the field of objects sources of the reflected EMW and transforming it to a set of the reflected R signals [80] (the Fig. 2.15).

For achievement of high characteristics of spatial resolution of objects the AS aperture characteristics (in particular, the complex ADP) should be coordinated with the corresponding characteristics of EMW reflected by MO. The following AS technical parameters significantly influence the structure of the accepted signals:

- the ADP width on azimuth $\Delta\alpha$ and elevation angle $\Delta\beta$;
- the amplification factor K_g;
- the average ADP LSL;
- the angular speed of scanning Ω_α in the set sector $\left(\Phi_\alpha, \Phi_\beta\right)$.

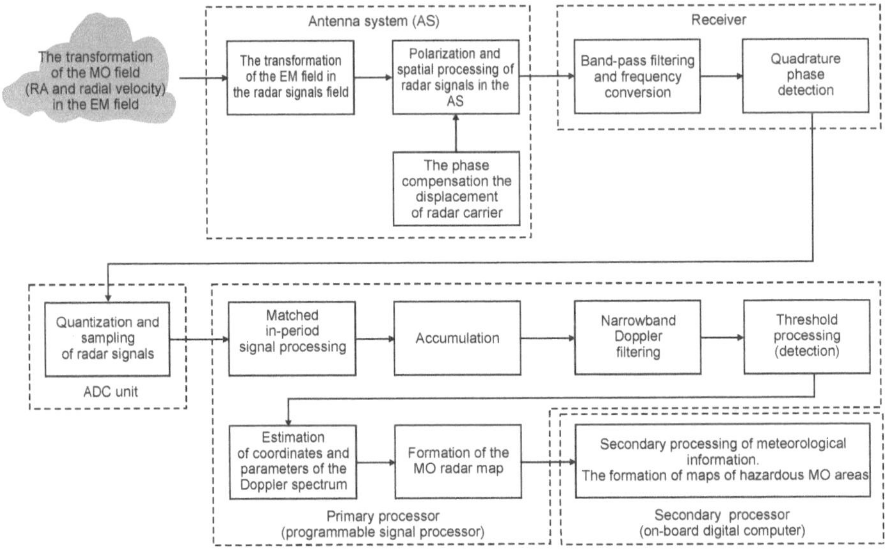

Fig. 2.15 A mathematical model of the signal processing board radar tract

The reflected R signals, as well as PS, have the form of a pack of radio impulses which coherent structure in case of spatially distributed target is defined by mutual movement of elementary reflectors in the allowed volume and also relative movement of all allowed volume in relation to the RS APC station in the course of reception.

Then the temporal (frequency) processing of signals is carried out. Partially it is carried out in the RS (frequency transformation, bandpass filtering, amplification, detecting). The receiver is linear with a necessary dynamic range (the Sect. 1.3) and complex transfer coefficient

$$K_i = |K_i| \exp\{j\varphi_i\},$$

where $|K_i|$ and φ_i—the module of transfer coefficient and phase invasion in the receiver. We believe that the phase invasion is a normal random value with the mathematical expectation $m_{\varphi i}$ and dispersion $\sigma_{\varphi i}^2$.

The bandwidth Δf_{IF} of the IFA of the receiver is coordinated with the width of the spectrum of the amplified signal of IF and, therefore, with the required interval of unambiguous determination of Doppler frequencies.

Besides, it is necessary to consider own noise of the RS receiver in the mathematical model of a path of processing. The standard model of complex readings of Gaussian white noise is used in the considered resolution element as the mathematical model of own noise of the airborne RS receiver at an input of the device of digital processing

$$\vec{N} = \left[n(1)\ n(2)\ \ldots\ n(M)\right]^T, \tag{2.121}$$

which AKM $\underline{\mathbf{B}}_N = \overline{\vec{N}^* \vec{N}^T}$ contains members of the type $b_{N\,in} = \sigma_N^2 \delta_{in}$, where σ_N^2—the noise dispersion determined by the noise coefficient K_n and the bandwidth Δf_{IF} of the receiver [4, 81] by the expression

$$\sigma_N^2 = kTav(K_n - 1)\Delta f_{IF}, \tag{2.122}$$

where $k = 1.38 \times 10^{-23} \mathrm{J}/\mathrm{K}$—the Boltzmann constant;
T_{av}—the absolute ambient temperature;
$\Delta f_{IF} = \tau_u^{-1}$—the bandwidth of the RS receiver determined by the PS duration.

Further, after allocation of quadrature components of complex envelope, the signals are transformed to a digital form by means of the ADC block. Here it should be noted that the path of processing of the reflected R signals is a source of not only thermal noise (in the receiver), but also noise of quantization of the ADC block. In this work the noise of quantization of ADC is not considered.

The received binary codes of discrete readings of the quadrature components of the accepted signals come to PSP carrying out operations of temporal processing (temporal selection, resolution, detection and assessment of parameters of the reflected signals, assessment of the impact of the movement of the RS carrier and its compensation) and also management of the systems of automatic adjustment of amplification of the RS receiver. The PSP is the specialized high-performance calculator in which architecture and command system are most adapted for the solution of problems of processing of signals [82, 83].

Thus, temporal processing of signals consists of carrying out a number of consecutive operations [72]:

1. The intra period processing of the accepted coherent signals including linear processing (amplification, frequency transformation, bandpass filtering, etc.) and the coordinated filtration or correlation processing.
2. The interperiod (Doppler) processing with the purpose to weaken negative impact of active and passive interferences.
3. The accumulation of signals and formation of some statistics about the accepted signals (decisive statistics) on the basis of which the decision on detection is made and the parameters of signals are estimated.
4. The threshold test of decisive statistics and realization of algorithms of detection and estimation of parameters of signals.

The operation 1 together with polarizing and spatial processing is the stage of intra period polarizing and space-time processing of single coherent signals. The operations 2…4 are the stage of interperiod processing of a set ("pack") of the signals reflected by each object in the course of uniform scanning by an ADP in the view zone or at repeated sounding of each direction of this zone.

The specifics of processing of signals in an airborne RS are, first of all, connected with the movement of its carrier. The movement of RS complicates the algorithms of formation of the R image since it is necessary to mate the separate fragments of the image shifted and turned from each other due to the movement, widens the spectrum of interperiod fluctuations of signals, etc. These negative effects cause the need for realization in the course of preprocessing of the phase correction of signals compensating influence of the movement and trajectory fluctuations of an AV on phase structure of signals.

Summing up the result, we will note the following:

1. In the mode of detection and assessment of degree of danger of MO zones there is the consecutive sequence of processing of R information on stages of primary and secondary processing. The view of space, selection, permission, detection and assessment of parameters of the signals reflected by MO for making decision on existence of MO zones with intensive air movement and definition of their relative position in the bound coordinate system belong to preprocessing (processing of signals). The assessment of the influence of the movement of the RS carrier and its compensation also belong to preprocessing. The secondary processing (processing of measurements) is made on the basis of the results of processing of signals, information on the current location and movement of the carrier for the purpose of formation of assessments of meteorological parameters of the MO areas. At the same time meteorological information is taken from spatial distributions (fields) of the measured radar values.

The mathematical model of a path of processing of R signals should include a set of algorithms of consistently carried out main functional operations of primary signal processing which involve the following objects of research:

- the algorithm of assessment of the DS parameters of the reflected signals which are unambiguously connected with parameters of the spectrum of velocities of MO particles;
- the algorithm of phase compensation of movement of the carrier.

2. The characteristics of a path of preprocessing of a RS as a uniform polarizing and space-time- filter should be agreed with the corresponding parameters of EMW, reflected by the allowed MO volume. In airborne RS with rather narrow-band probing fluctuations and small AS apertures the polarizing and space-time processing of signals is practically always a separable, i.e. divisible on polarizing, spatial and temporal.

Polarizing and spatial filtration of the field of the reflected EMW and its transformation into a set of the reflected R is carried out by the RS AS. Temporal (frequency) processing of signals is partially carried out in an analog form in the RS linear receiver (frequency transformation, bandpass filtering, amplification, detecting, coordinated intra period processing), partially—in a digital form in RSP (temporal selection, resolution, detection and assessment of parameters of the reflected signals, assessment of the influence of the movement of the RS carrier and its compensation).

The Main Conclusions for Sect. 2.2

On the basis of the results of the second section it is possible to make the following main conclusions:

1. The aggregate mathematical model of the problem of detection and assessment of the degree of danger of MO should be built with use of the known principles of block construction and specialization of mathematical models of radio engineering systems and include three main parts:

   – the model of formation of R signals (MO model, model of movement of the RS carrier, model of a useful signal);
   – the model of a path of processing of signals of airborne RS;
   – the block of programs of control and processing of the results of modelling.

2. Proceeding from the purpose and objectives of the present work, modelling of the developed algorithms of digital processing of signals in airborne RS should be carried out at the functional level. As the majority of modern devices of digital processing of signals (both reprogrammed and with unconditional logic) operates with digital readings of quadrature components of the complex envelope of the accepted signals, the method of complex envelope adequate to this representation is used in the developed model.
3. The MO model should consider the spatial distribution of its physical parameters (radial wind speed in the conditions of WS and turbulence, water content). For detection and assessment of dangers of WS areas located at rather big height the model of the wind speed field in the descending flow in the form of "a ring vortex" is preferable.
4. The spatial movement of radar station represents superposition of the movement of the carrier on a fixed basic trajectory and random deviations from a basic trajectory. APC deviation from a basic trajectory is defined by TI, the rotation of the fuselage around CM, flexural fluctuations and torsion of an AV structure and also aerodynamic vibration. All specified fluctuations have narrow-band character. In this case the model of an AV movement in the turbulent atmosphere can be presented by the system of the linear differential equations with constant coefficients for each of coordinates.
5. The signal reflected by the allowed MO volume on an input of RSP can be described by discrete readings of quadrature components of the complex envelope, representing two statistically connected discrete random processes. If the radial component of average wind speed $\bar{V} = 0$, then the specified processes can be considered as mutually independent Gaussian random processes. In this case the mathematical model of the signal reflected by MO represents the model of two independent Markov processes of the first order. The nonzero radial component of average wind speed leads to emergence of correlation between these random processes, and the module of coefficient of mutual correlation depends on the value $\bar{V}$, and the sign—on the sign (direction) of speed. At the same time the quadrature components of the complex envelope the reflected signal become the correlated Markov processes of the second order.
6. The spatial distributions of MO specific EER, total EER and average power of the signal reflected by the allowed volume are defined on the basis of distribution of RR. The carried-out calculation showed that total EER of the allowed MO

volume is very considerable (tens-hundreds of sq.m) and increases under the square law with increase in range.

7. The DS of the reflected MO signal corresponds to the spectrum of radial velocities of separate reflectors taking into account the contribution of each of them to power of the reflected signal, i.e. at the analysis of the spatial distribution of radial velocities of reflectors on the MO volume it is necessary to consider heterogeneity of the RR spatial field (specific EER) of MO. However, in the majority of practical cases, especially for airborne RS with high resolution, it is possible to consider the fields of radial velocity and specific EER as locally stationary within the V_i volume.
8. The time limitation of R contact of an airborne RS with MO does not allow to receive the reflected signal of large volume and by that to provide the required accuracy of measurement and resolution of a radar station on the Doppler frequency (velocity). The packed nature of the reflected signal leads to emergence in its spectrum at the expense of Gibbs phenomenon of side lobes with the maximum level near −13 dB. However, the assumption of equality to zero of the MO signal outside the interval of observation is not true as the signal is narrow-band, and its ACF is close to periodic one. The target situation causing formation of the MO signal can be rather precisely simulated by the linear system of a finite order (the AR model). In this case for the assessment of the DS parameters of the reflected signal it is necessary to define the values of AR coefficients and dispersion of the forming (excitation) signal.
   The spectra of signals approximated by AR models have very simple physical interpretation: each AR coefficient corresponds to existence in the reflected signal of the narrow-band random sequence with its average frequency and spectrum width. This takes place in case MO consists of several groups of the reflectors moving with various velocities. In real conditions the signals reflected by MO can be adequately presented by a random process of a low (second or fifth) order.
9. The procedure of processing of R information in the mode of the detection and assessment of danger of MO zones consists of consistently performed stages of primary and secondary processing. The view of space, selection, permission, detection and assessment of parameters of the signals reflected by MO for making decision on existence of dangerous zones of MO and definition of their relative position in the BCS belong to preprocessing (processing of signals). The assessment of the influence of the movement of the RS carrier and its compensation also belong to preprocessing. The secondary processing (data processing) is made on the basis of the results of processing of signals, information on the current location and movement of the carrier for the purpose of formation of assessments of meteorological parameters of the MO areas. At the same time meteorological information is taken from spatial distributions (fields) of the measured R values.

The mathematical model of a path of processing of R signals should include a set of algorithms of consistently carried out main functional operations which involve the following objects of research:

- the algorithm of assessment of the DS parameters of the reflected signals which are unambiguously connected with parameters of the spectrum of velocities of MO particles;
- the algorithm of phase compensation of movement of the carrier.

References

1. Melnik Yu. A., Stogov G. V. Bases of radio engineering and radio engineering devices. - M.: "Sov. Radio", 1973. - 368 pages.
2. Mikhaylutsa K.T. Digital processing of signals in airborne radar-tracking systems: Text book/Eds. V.F. Volobuyev. - L.: LIAP, 1988. - 78 pages.
3. Alyanakh I.N. Modelling of computing systems. - L.: Mechanical engineering, 1988. - 233 pages.
4. Borisov Yu.P. Mathematical modelling of radio systems: Ed. book for higher education institutions. - M.: "Sov. radio", 1976. - 296 pages.
5. Modelling in radar-location/A.I. Leonov, V.N. Vasenev, Yu.I. Gaydukov, et al.; Eds. A.I. Leonov. - M.: Soviet radio, 1979. - 264 pages.
6. Matveev L.T. Course of the general meteorology. Physics of the atmosphere. - L.: Hydrometeoizdat, 1984. - 751 pages.
7. Aviation meteorology: Textbook/A.M. Baranov, O.G. Bogatkin, V.F. Goverdovsky, et al..; Eds. A.A. Vasilyev. - SPb.: Hydrometeoizdat, 1992. - 347 pages.
8. Vasin I.F. Influence of wind shear on safety of flights of aircrafts. - In the book: Pub. house VINITI. Results of science and technology. - M.: Air transport, 1980. - V. 8, p. 5–30.
9. RDR-4B. Forward Looking Windshear Detection/ Weather Radar System. User's Manual with Radar Operating Guidelines. Rev. 4/00. - Honeywell International Inc., Redmond, Washington USA, 2000.
10. Fujita T.T., Byers H.R. Spearhead echo and downburst in the crash of an airliner. Mon. Wea. Rev., 1977, №105, p. 1292–1346
11. Ivan M. A ring-vortex downburst model for flight simulations. Journal of Aircraft, 1986, v. 23, 13, p. 232–236.
12. Kalinin A.V., Monakov A.A. Assessment of parameters of the meteophenomena, dangerous to aircraft, by a radar method: Report. - In the book: The fourth scientific session of graduate students of SUAI devoted to the World day of aircraft and astronautics and the 60 anniversary of SUAI (St. Petersburg, March 26–30, 2001): Collection of reports. P.1. Technical sciences. - SPb.: publishing house of St. Petersburg State University of Aerospace Instrumentation, 2001. - p. 236–238.
13. Holmes J.D., Oliver S.E. An empirical model of a downburst. Engineering Structures, 2000, v. 22, 19, 1167–1172.
14. Doviak R., Zrnich. Doppler radars and meteorological observations: The translation from English - L.: Hydrometeoizdat, 1988. - 511 pages.
15. Taylor G. The loadings applied to a plane. - M.: Mechanical engineering, 1971.
16. Atlas D. Achievements of radar meteorology/ Translation from English - L.: Hydrometeoizdat, 1967. - 194 pages.
17. Tatarsky V.I. The theory of the fluctuation phenomena at distribution of waves in the turbulent atmosphere. - M.: Publishing house of Academy of Sciences of the USSR, 1959. - 232 pages.
18. Dobrolensky Yu.P. Dynamics of flight in the rough air. - M.: Mechanical engineering, 1969.
19. Zubkov B.V., Minayev E.R. Bases of safety of flights. - M.: Transport, 1987. - 143 pages.
20. Banakh V.A., Verner X., Smalikho I.N. Sounding of turbulence of clear sky by a doppler lidar. Numerical modelling. Optics of the atmosphere and ocean, 2001, v. 14, No. 10, pages 932–939.

21. Radar stations with digital synthesizing of an antenna aperture. V.N. Antipov, V.T. Goryainov, A.N. Kulin et al.; Eds. V.T. Goryainov. - M.: Radio and Communication, 1988. - 304 pages.
22. Problems of creation and application of mathematical models in aircraft. - M.: Science, 1983 - (Ser. "Cybernetics issues").
23. Introduction to aero autoelasticity. S.M. Belotserkovsky et al. - M.: Science, 1980 – 384 pages.
24. Vorobiev V.G., Kuznetsov S.V. Automatic flight control of planes: Ed. book for higher education institutions. - M.: Transport, 1995. - 448 pages.
25. Flight operation safety / Eds. R.V. Sakach. - M.: Transport, 1989. - 239 pages.
26. Krasyuk N.P., Rosenberg V.I. Ship radar-location and meteorology. - L.: Shipbuilding, 1970. - 325 pages.
27. Levin B.R. Theoretical bases of statistical radio engineering. - M.: Radio and Communication, 1989. - 656 pages.
28. Tikhonov V.I. Statistical radio engineering. - M.: Radio and Communication, 1982. - 624 pages.
29. Borisov Yu.P., Tsvetnov V.V. Mathematical modelling of radio engineering systems and devices. - M.: Radio and Communication, 1985. - 176 pages.
30. Multipurpose radar complexes of fighter planes. V.N. Antipov, S.A. Isaev, A.A. Lavrov, V.I. Merkulov; Eds. G.S. Kondratenkov. - M.: Voyenizdat, 1994.
31. Stepanenko V. D. Radar-location in meteorology. - the 2nd ed. - L.: Hydrometeoizdat, 1973. - 343 pages.
32. Tikhonov V.I., Mironov M.A. Markov processes. - M.: Sov. Radio, 1977. - 488 pages.
33. Chernyshov E.E., Mikhaylutsa K.T., Vereshchagin A.V. Comparative analysis of radar methods of assessment of spectral characteristics of moisture targets: The report on the XVII All-Russian symposium "Radar research of environments" (20–22.04.1999). - In the book: Works of the XVI-XIX All-Russian symposiums "Radar research of environments". Issue 2. - SPb.: VIKU, 2002 - p. 228–239.
34. Feldman Yu.I., Mandurovsky I.A. The theory of fluctuations of the locational signals reflected by the distributed targets. - M.: Radio and Communication, 1988. - 272 pages.
35. Krasyuk N.P., Koblov V.L., Krasyuk V.N. Influence of the troposphere and the underlying surface on work of a radar station. - M.: Radio and Communication, 1988. - 216 pages.
36. Radar methods of a research of the Earth/Eds. Yu.A. Melnik. - M.: Sov. radio, 1980. - 264 pages.
37. Ivanov, Yu.V. Asymptotic efficiency of algorithms of space-time processing in coherent and pulse radar stations. Ed.house of mag. Radio electronics (News of Higher Educational Institutions) - Kiev, 1990. - 12 pages (Dep. in VINITI 06.12.90 No. 6151-B90).
38. Ryzhkov A.V. Meteorological objects and their radar characteristics. Foreign radio electronics, 1993, No. 4, pages 6–18.
39. Mazin I. P., Hrgian A.H. Clouds and cloudy atmosphere: Reference book. - L.: Hydrometeoizdat, 1989. - 647 pages.
40. Rosenberg I.V. Dispersion and weakening of electromagnetic radiation by atmospheric particles. - L.: Hydrometeoizdat, 1972. - 348 pages.
41. Brylev G.B., Gashina S.B., Nizdoyminoga G.L. Radar characteristics of clouds and rainfall. - L.: Hydrometeoizdat, 1986. - 231 pages.
42. Mikhaylutsa K.T., Chernulich V.V. The device of digital processing of signals for the incoherent weather radar. Issues of special radio electronics. Ser. RLT, 1981, No. 20.
43. Aviation radar-location: Reference book. Eds. P. S. Davydov. - M.: Transport, 1984. - 223 pages
44. Ivanov A.A., Melnichuk Yu.V., Morgoyev A.K. The technique of assessment of vertical velocities of air movements in heavy cumulus clouds by means of the doppler radar. Works of the CAO, 1979, issue 135, pages 3–13.
45. Bracalente E.M., Britt C.L., Jones W.R. Airborne Doppler Radar Detection of Low-Altitude Wind Shear. Journal of Aircraft, 1990, Vol. 27, 12, p.p. 151–157.
46. Okhrimenko A.E. Bases of radar-location and radio-electronic fight: Text book for higher education institutions. P.1. Bases of radar-location. - M.: Voyenizdat, 1983. - 456 pages.
47. Theoretical bases of a radar-location: Ed. book for higher education institutions. A.A. Korostelev, N.F. Klyuev, Yu.A. Melnik, et al.; Eds. V.E. Dulevich. - M.: Sov. Radio, 1978. - 608 pages.

48. The airborne radar for measurement of velocities of vertical movements of lenses in clouds and rainfall. V.M. Vostrenkov, V.V. Ermakov, V.A. Kapitanov et al. Works of the CAO, 1979, issue 135, pages 14–23.
49. The doppler radar of vertical sounding for a research of dynamic characteristics of cloud cover from a plane board. V.M. Vostrenkov, V.V. Ermakov, V.A. Kapitanov et al. Works of the 5th All-Union meeting on radar meteorology. - M.: Hydrometeoizdat, 1981. - p. 145–148.
50. Smith P., Hardy K., Glover K. Radars in meteorology// TIIER, t.62, No. 6, 1974, pages 86–112.
51. Meister Yu.L., Piza D.M. Weight functions for DFT in the systems of reception and processing of radar signals. "Radioyelektronika. Informatika. Upravlinnya" (Zaporizhia, ZGTU), 2001, No. 1, pages 8–12.
52. Mironov M.A. Assessment of parameters of the model of autoregression and moving average on experimental data. Radio engineering, 2001, No. 10, pages 8–12.
53. Gong S., Rao D., Arun K. Spectral analysis: from usual methods to methods with high resolution. - In the book: Superbig integrated circuits and modern processing of signals/ Eds. Gong S., Whitehouse H., Kaylat T.; the translation from English eds. V.A. Leksachenko. - M.: Radio and Communication, 1989. - P. 45–64.
54. Marple Jr. S.L. Digital spectral analysis and its applications: The translation from English - M.: Mir, 1990. - 584 pages.
55. Bakulev P.A., Koshelev V.I., Andreyev V.G. Optimization of ARMA-modelling of echo signals. Radio electronics, 1994, No. 9, pages 3–8. (News of Higher Educational Institutions).
56. Bakulev P.A., Stepin V.M. Features of processing of signals in modern view RS: Review. Radio electronics, 1986, No. 4, pages 4–20. (News of Higher Educational Institutions).
57. Khaykin S., Carry B.U., Kessler S.B. The spectral analysis of the radar disturbing reflections by the method of the maximum entropy. TIIER, 1982, v. 70, No. 9, pages 51–62
58. Voyevodin V.V., Kuznetsov Yu. A. Matrixes and calculations. – M.: Science, 1984 – 320 pages.
59. Vostrenkov V.M., Ivanov A.A., Pinskiy M.B. Application of methods of adaptive filtration in the doppler meteorological radar-location. Meteorology and hydrology, 1989, No. 10, pages 114–119.
60. Hovanova N.A., Hovanov I.A. The methods of the time series analysis. - Saratov: Publishing house of GosUNTs "College", 2001. - 120 pages.
61. Koshelev V.I., Andreyev V.G. Optimization of AR models of processes with a polymodal spectrum. Radio electronics, 1996, No. 5, pages 43–48. (News of Higher Educational Institutions).
62. Koshelev V.I., Andreyev V.G. Application of ARMA-models at modelling of echo signals. Radio electronics, 1993, No. 7, pages 8–13. (News of Higher Educational Institutions).
63. Andreyev V.G., Koshelev V.I., Loginov S.N. Algorithms and means of the spectral analysis of signals with a big dynamic range. Radio electronics issues Ser. RLT. 2002, issue 1–2, pages 77–89.
64. Yermolaev V.T., Maltsev A.A., Rodyushkin K.V. Statistical characteristics of AIC, MDL criteria in a problem of detection of multidimensional signals in case of short selection: Report. – In the book: The third International conference "Digital Processing of Signals and Its Application": Reports. – M.: NTORES, 2000 - Volume 1. P. 102–105.
65. Ulrych T.J., Clayton R.W. Time Series Modeling and Maximum Entropy. Phys. Earth Planet. Inter., V. 12, p.p. 188–200, August 1976.
66. Ulrych T.J., Ooe M. Autoregressive and Mixed ARMA Models and Spectra. - In the book: Nonlinear Methods of Spectral Analysis, 2nd ed., S. Haykin ed., Springer-Verlag, New York, 1983
67. Manual on flights in CA of the USSR (NPP CA - 85). - M.: Air transport, 1985. - 254 pages.
68. Radar systems of air vehicles: Textbook for higher education institutions/ Eds. P. S. Davydov. - M.: Transport, 1977. - 352 pages.
69. Tuchkov N.T. The automated systems and radio-electronic facilities of air traffic control. - M.: Transport, 1994. - 368 pages.
70. ARINC-708. Airborne Weather Radar. - Aeronautical Radio Inc., Annapolis, Maryland, USA, 1979.

71. ARINC-708A. Airborne Weather Radar with Forward Looking Windshear Detection Capability. - Aeronautical Radio Inc., Annapolis, Maryland, USA, 1993.
72. Kuzmin S. Z. Bases of design of systems of digital processing of radar processing. - M.: Radio and Communication, 1986. - 352 pages.
73. Radar stations of the view of Earth. G.S. Kondratenkov, V.A. Potekhin, A.P. Reutov, Yu.A. Feoktistov; Eds. G.S. Kondratenkov. - M.: Radio and Communication, 1983. - 272 pages.
74. Active phased antenna arrays. V.L. Gostyukhin, V.N. Trusov, K.G. Klimachev, Yu.S. Danich; Eds. V.L. Gostyukhin. - M.: Radio and Communication, 1993. - 272 pages.
75. Antennas and UHF devices. Design of the phased antenna arrays. Eds. D.I. Voskresensky. - M.: Radio and Communication, 1994. - 592 pages.
76. Processing of signals in multichannel RS. A.P. Lukoshkin, S.S. Korinsky, A.A. Shatalov, et al.; Eds. A.P. Lukoshkin. - M.: Radio and Communication, 1983. - 328 pages.
77. Cook Ch., Burnfield M. Radar signals: The translation from English - M.: Sov. radio, 1971. - 568 pages.
78. Svistov V.M. Radar signals and their processing. - M.: Sov. Radio, 1977. - 448 pages.
79. Protection of radar-tracking systems against interferences. State and trends of development. Eds. A.I. Kanashchenkov and V.I. Merkulov. - M.: Radio engineering, 2003. - 416 pages.
80. Korostelev A.A. Space-time theory of radio systems: Text book. - M.: Radio and Communication, 1987. - 320 pages.
81. Bykov V.V. Digital modelling in statistical radio engineering. - M.: Sov. radio, 1971. - 328 pages.
82. Mikhaylutsa K.T., Ushakov V.N., Chernyshov E.E. Processors of signals of aerospace radio systems. - SPb.: JSC Radioavionika, 1997. - 207 pages.
83. Multipurpose pulse-doppler radar stations for fighter planes: Review. - M.: RDE NIIAS, 1987. - 70 pages.

Chapter 3
The Methods and Algorithms of Processing of Signals Improving Observability and Accuracy of the Assessment of Parameters of Meteorological Objects of Airborne Radar Stations

A big variety of conditions of formation of the signals reflected by MO causes essential aprioristic uncertainty of their parameters that does not give the opportunity of use of any known in advance characteristics for the organization of optimum processing of these signals, for example, by Bayesian methods or by the methods of maximum of posteriori density of probability. Besides, the feature of MO signals is spatial heterogeneity and temporal not stationarity of their parameters. In this regard, we will look for the method of processing of the signals reflected by MO in the class of the methods which are characteristic for the work in the conditions of aprioristic uncertainty that is among the adaptive methods.

All methods of processing of the signals reflected by MO can be conditionally broken into two classes: the methods of processing of blocks of data and the methods of processing of consecutive data [1]. The first is intended for processing of blocks of the accumulated readings of data of some fixed volume (packs). It is expedient to apply similar methods when the volume of the available pack is strongly limited, but it is desirable to receive the assessment with the best possible characteristics. In the presence of longer selections it is possible to use the methods of consecutive estimation providing updating of the received assessments of parameters in process of receipt of each new reading.

At the solution of the problem of detection and assessment of danger of MO zones by means of airborne RS the preference should be given to methods of the first class as, firstly, the time of R observation is limited owing to the high speed of an AV flight and big range of change of the MO specific EER, and secondly, updating of the received assessments of parameters of the spectrum of MO velocities with the frequency of repetition of PS demands carrying out on each step of large volume of calculations exceeding possibilities of the existing and perspective RSP.

On the other hand, the methods of the spectral analysis of the random signals reflected by MO can be divided into two groups on other criterion of classification: nonparametric and parametric methods. In nonparametric methods only the information contained in reading of the analysed signal is used. Parametric methods assume existence of some statistical model of a random signal and process of the spectral analysis in this case includes determination of parameters of this model.

Vereshchagin A. V. et al., *Signal Processing of Airborne Radar Stations*,
Springer Aerospace Technology, https://doi.org/10.1007/978-981-13-9988-6_3

3.1 The Methods and Algorithms of Processing of Signals of Airborne Meteo RS for the Assessment of Frequency and Width of a MO Doppler Spectrum

The first step at the assessment of the average frequency and RMS width of a DS is the normalization of the accepted signals on power (according to the (1.17) and (1.18)). The assessment of the average power of signals (1.12) is made by its averaging on M consecutive readings of P_{cm} (2.59). In practice for the assessment of the average power the readings not of power, but of a signal from the output of the Q_m receiver which are functionally connected with the value of P_{cm} are averaged. In this regard the expression (2.62) can be presented in the form

$$\overline{P}_c \sim \overline{Q} = \frac{1}{M}\sum_{m=1}^{M} Q_m. \tag{3.1}$$

The dependence between P_{cm} and Q_m is defined by the type of amplitude characteristic (AC) of the receiver. The receiver AC can be of the following types:

- linear AC, characterized by dependence $Q_m \sim \sqrt{P_{cm}}$;
- square AC, at which $Q_m \sim P_{cm}$;
- logarithmic AC, at which $Q_m \sim \log P_{cm}$.

The envelope of the reflected signal for each allowed volume is exposed to analog-digital transformation and in the form of the code comes to RSP carrying out the accumulation of the set number of readings (2.32) of the envelope of the signal for receiving of assessments of average power.

3.1.1 *Nonparametric Methods of the Assessment of Parameters of a Doppler Spectrum of the Signal Reflected by MO*

The nonparametric methods of spectral estimation which are traditionally applied in meteorological radar-location are based on calculation of spectral density of power on one of two equivalent formulas [2]

$$S(f) = \sum_{m=-\infty}^{\infty} B(mT_n)e^{-j2\pi f m T_n}, \tag{3.2}$$

$$S(f) = \lim_{M\to\infty}\frac{1}{M}\left|\sum_{m=0}^{M-1} s(mT_n)e^{-j2\pi f m T_n}\right|^2, \tag{3.3}$$

where $B(mT_n)$—the ACF of the signal $s(m\ T_n)$.

The formula (3.2) is based on the Blackman and Tukey approach, consisting in preliminary receiving of the assessment of ACF on a final interval and the subsequent application of the Fourier transformation.

The assessment of ACF $B(mT_n)$ of a signal can be received in the following form

$$\hat{B}(mT_n) = \frac{1}{M-m}\sum_{n=0}^{M-m-1} s([n+m]T_n)s^*(nT_n).$$

When using the Blackman and Tukey approach AKF is supposed to be equal to zero outside the observation interval

$$S(f) = \sum_{m=0}^{M-1} \hat{B}(mT_n)e^{-j2\pi f mT_n}. \qquad (3.4)$$

This assumption leads to emergence of side lobes in the spectrum of the processed signal at the expense of Gibbs phenomenon. For smoothing of transition to zero values before Fourier transformation ACF is often exposed to weight processing

$$S(f) = \frac{1}{P_w}\sum_{m=0}^{M-1} \hat{B}(mT_n)w(mT_n)e^{-j2\pi f mT_n}. \qquad (3.5)$$

where $w(m\ T_n)$—the function of a weight window;

$$P_w = \frac{1}{M}\sum_{m=0}^{M-1} |w(mT_n)|^2.$$

Application of the weight processing directed to reduction of the masking action of side lobes leads to power losses and also to the need of allocation of temporal resources on weight processing.

The methods based on calculation of ACF received the name of correlogram methods of spectral assessment [3]. The optimum correlogram method of assessment of the DS parameters is the maximum likelihood method (ML) providing the assessment with the minimum dispersion.

3.1.1.1 The Method of Maximum Likelihood

According to the theory of statistical decisions, the operation of measurement of the not power parameter A of a signal on the ML method in the optimum processing device consists in formation of the correlation integral [4–6]

$$Z(A) = \int_{0}^{T} x(t)s_0^*(t, A)dt \tag{3.6}$$

and finding of the value of the parameter A corresponding to a maximum of this integral.

In (3.6) $x(t) = s(t) + n(t)$—the input mixture of a useful signal and interferences, $s_0(t, A)$—the basic signal of the receiver, T—the analysis time. The device constructed taking into account (3.6) is not nontracking measuring instrument.

As usually the values of interferences and the parameter A are unknown in advance, the optimum processing should be carried out for all set of discrete values of A. The measuring instrument can be executed in the form of the multichannel (parallel) device, each of channels of which is set for the value of the measured parameter and calculates for it the value of correlation integral $Z(A_i), i = \overline{1, \ldots, n}$, or in the form of the analyser which is consistently adjusted on various values of the parameter. The number of channels n of the parallel analyser is determined by the relation of a priori interval of values of the parameter A to the required resolution ΔA on this parameter. The decision on the assessment of the parameter A is made on the basis of the choice of the channel of the processing device with the maximum output signal (3.6).

At the assessment of average frequency $\bar{f}$ of the DS signals reflected by DS, the number of channels of the optimum multichannel device of assessment is defined by the required accuracy of its measurement $\delta \bar{f}$

$$n = (\bar{f}_{\max} - \bar{f}_{\min})/\delta \bar{f},$$

where $\bar{f}_{\min} - \bar{f}_{\max}$—the priori interval of possible values $\bar{f}$.

Performance of the listed operations for receiving of the assessments of the DS parameters by the ML method demands the large volume of calculations and considerable time for receiving of necessary volume of selection of the reflected signal, especially in the conditions of spatial heterogeneity of the estimated parameters. In case of use of land radar stations it is still possible to execute all volume of calculations for receiving of ML assessments, but at installation of radar station on a board of a moving carrier it is almost not possible to solve the objective in real time by the ML method.

Now simpler suboptimum methods of assessment of the DS parameters meeting the following requirements are widely spread:

- ensuring work in the conditions of aprioristic uncertainty about parameters of the processed signals;
- ensuring work in real time;
- whenever possible, the single-channel realization applicable in airborne radar stations.

One of similar methods is the modified ML method based on replacement of the correlation integral (3.6) with the autocovariant integral

$$Z'(\tau) = \int_0^T s^*(t)s(t+\tau)dt. \tag{3.7}$$

The condition of applicability of the expression (3.7) is the requirement

$$\Delta f_{\text{if}} >> \bar{f}, \tag{3.8}$$

where Δf_{if}—the IFA bandwidth.

In the coherent and pulse radar stations providing high resolution on range due to sounding by short pulse signals, the IFA bandwidth is out of the condition $\Delta f_{\text{if}} \approx K/\tau_u$, where K = 0.5–2—the coefficient; τ_u—the PS duration. At duration τ_u within 0.1–1 μs the IFA bandwidth is within 3–130 MHz thanks to what correctness of the inequality (3.8) is always provided.

In case of realization of the suboptimum algorithm (3.7) it is necessary to estimate the ACM $\underline{\hat{\mathbf{B}}}$ and to make its inversion. At the same time it is important to provide independence of readings of $s(t)$ and $s(t + \tau)$ at all values of a delay τ.

The assessment of the average frequency $\bar{f}$ corresponds to the minimum of the square form of the type

$$J(f = \bar{f}) = \min\left\{\sum_{i,k} x^*(i)b'_{i-k}x(k)\exp[j2\pi f(i-k)T_n]\right\}. \tag{3.9}$$

where b'_{i-k}—the elements of the inversed autocovariant matrix $\underline{\hat{\mathbf{B}}}^{-1}$.

Thus, at using the modified ML method it is necessary to perform the following operations [7]:

- п—to make the assessment of the ACM of the reflected signal on a pack of readings with the length $M_E \geq 10\,\text{Ent}\{\tau_K/T_n\}$, where τ_K—the interval of correlation of the reflected signal;
- to make the inversion of the received ACM;
- to calculate the assessment $\bar{f}$ by minimization of the square form $J(f)$ (3.9).

The assessment of the width of the spectrum Δf is made in the same way.

Minimization of the square form $J(f)$ (3.9) on all possible values of Doppler frequencies and also the inversion of the ACM of the reflected signal demand essential time and hardware consumption. As a result the requirements to the device of suboptimum processing given above are not fulfilled.

In land meteo radar stations with coherent processing of signals the suboptimum methods of direct assessment of the moments of DS without transformation of signals to frequency area demanding the significantly smaller volume of the carried-out calculations and the used memory volume are used. They are based on the assumption that statistics of the signal disseminated on set of the randomly located in space HM is Gaussian [8] owing to what DS of MO signals is also considered to be Gaussian. It is promoted by the forms of the main lobe of DP of most radar station close to Gaussian function and also the distribution of velocities of HM. The Gaussian spectrum is defined completely by the average frequency $\bar{f}$ and the RMS width Δf.

3.1.1.2 The Autocovariant Method (the Method of "Paired Pulses")

The most known of suboptimum methods is the autocovariant method (the method of "paired pulses") [9–11] based on the assessment of derivatives of the complex ACF of the reflected signals at a zero delay corresponding according to the theorem of the moments to the moments of spectral density

$$\bar{f} = \frac{1}{j2\pi}\frac{B'(\tau)}{B(0)}, \tag{3.10}$$

$$\overline{\Delta f^2} = -\left(\frac{1}{2\pi}\right)^2 \frac{B''(\tau)}{B(0)} = \Delta f^2 + \bar{f}^2, \tag{3.11}$$

where $B'(\tau)$ and $B''(\tau)$—the first and second derivative of the ACF on the argument τ.

It follows from the expressions (3.10) and (3.11) that

$$\Delta f = \frac{1}{2\pi}\left\{-\frac{B''(\tau)}{B(0)} + \left[\frac{B'(\tau)}{B(0)}\right]^2\right\}^{1/2}. \tag{3.12}$$

The ACF of the signal reflected by MO and having Gaussian distribution is equal to

$$B(\tau) = \bar{P}_c \exp\left\{j2\pi\bar{f}\tau - 2\pi^2\Delta f^2\tau^2\right\}, \tag{3.13}$$

where $\bar{P}_c = B(0)$—the average power of the reflected signal.

Considering (3.13), the expressions (3.10) and (3.12) are transformed to the form [9, 11, 12]

$$\bar{f} = \frac{1}{2\pi\tau}\mathrm{Arg}[B(\tau)] = \frac{1}{2\pi\tau}\mathrm{arctg}\frac{\mathrm{Im}[B(\tau)]}{\mathrm{Re}[B(\tau)]}, \tag{3.14}$$

$$\Delta f = \left\{-\frac{1}{2(\pi\tau)^2}\ln\left[\frac{|B(\tau)|}{\bar{P}_c}\right]\right\}^{1/2}. \tag{3.15}$$

Otherwise, the argument of the ACF is not displaced assessment of the first moment of DS.

As shown in [9], after expanding the logarithm in (3.15) to the series near zero delays, it is possible to receive the assessment Δf, suitable for hardware realization

$$\Delta f = \left\{ \frac{1}{2(\pi\tau)^2} \left| 1 - \frac{|B(\tau)|}{\bar{P}_c} \right| \right\}^{1/2}. \tag{3.16}$$

The assessment (3.16) for wide spectra is asymptotically displaced, unlike the assessment (3.15).

3.1.1.3 The Periodogram Methods of Spectral Estimation

The second approach to the assessment of spectral characteristics (according to (3.3)) is based on Wiener-Khintchine [13] theorem for ergodic process. The similar methods consisting of direct data transformation and the subsequent averaging of the received assessments, are called periodogram methods of spectral estimation. Periodogram is the power spectrum assessment defined as the square of Fourier transformation from the block of data of finite length [14]

$$S(f) = \frac{1}{M} \left| \sum_{m=0}^{M-1} s(mT_n) e^{-j2\pi f m T_n} \right|^2. \tag{3.18}$$

Periodogram with the accuracy to constant multiplier coincides with the assessment by the formula (3.3).

The periodogram methods of spectral estimation are widely spread due to the emergence of the FFT effective procedures and the subsequent reduction in cost of the corresponding hardware. For calculation of the average frequency of DS the periodogram (3.18) is firstly calculated, then the rough assessment of average frequency k_m/MT_p where the index of the maximum spectral component is chosen as the k_m index is defined. Then the assessment of the average frequency has the form

$$\bar{f} = \frac{1}{MT_n} \left\{ k_m + \frac{k_m}{\hat{P}} \sum_{k=k_m-M/2}^{k_m+M/2} (k - k_m) \hat{S}\left(\frac{\mathrm{mod}_M k}{T_n} \right) \right\}, \tag{3.19}$$

where $\hat{P}$—the power, total on a periodogram;
$mod_M\ k$—the rest from integer division of k by M;
$\hat{S}(\ldots)$—the value of the *k*-th a spectral component.

The assessment of the spectrum RMS by this method is defined by the expression

$$\Delta f = \left(\frac{1}{\hat{P} T_n^2} \sum_{k=k_m-M/2}^{k_m+M/2} \left[\frac{k}{M} - \bar{f} T_n \right]^2 \hat{S}\left(\frac{\mathrm{mod}_M k}{T_n} \right) \right)^{1/2}. \tag{3.20}$$

This assessment has shift if it is comparable with the Nyquist interval ($1/T_p$). In this aspect the method of paired pulses is more preferable in comparison with the periodogram method, but in case of narrow spectra the shift of assessment is insignificant.

The advantage of the periodogram method is existence of assessment of spectral density of power of the reflected signals for each allowed volume that allows to perform, besides the assessments of the spectral moments, other operations of Doppler processing (suppression of US, etc.).

The analysis of computing expenses for the spectrum assessment by the periodogram method shows that it is required to execute about $N \log_2 N$ complex operations, and the complexity of the method of paired pulses is proportional to N [7, 9].

For increase in jam resistance of the assessment of spectrum parameters, that is reduction of their dependence on turbulences and splashes at remote frequencies, the Welch [3, 15] method which is modification of the classical periodogram method and consisting in consecutive performance of the following actions is widely applied in practice of the applied spectral analysis:

1. Breaking of a pack of the accepted reflected signals $s(mT_n)$ of the length M on the finite number N of not overlapping or partially overlapping intervals $s^{(n)}(mT_n) = s[(m + (n-1)M_C))T_n]$ with the length of $\mathrm{M_n} < \mathrm{M}$. Here $n = 1, \ldots, N$, $m = 0, \ldots, M_n - 1$, $\mathrm{N} = \mathrm{int}\{(M - M_n)/[(1-C)M_n]\} + 1$, $M_C = M_n - \mathrm{int}\{M_n C\}$, $C < 1$—the overlapping degree.
2. Centring of a signal on each site: $s^{(n)}(mT_n) = s^{(n)}(mT_n) - E_n\{s(mT_n)\}$, where $E_n\{s(mT_n)\} = \frac{1}{M_n}\sum\limits_{m=0}^{M_n-1} s^{(n)}(mT_n)$—the average value of a signal on the n-th site.
3. Weighing of a signal by a window function and calculation of energy of a window

$$s_w^{(n)}(mT_n) = w(mT_n)s^{(n)}(mT_n), \qquad m = 0, \ldots, M_n - 1,$$
$$P_w = \frac{1}{M}\sum_{m=0}^{M-1} |w(mT_n)|^2.$$

4. The periodogram is calculated for each n-th site of breaking with use of the FFT algorithm $S_n(\omega)$. The assessment of PSD is formed by averaging on number of intervals of periodogram values

$$S(f) = \frac{1}{N}\sum_{n=1}^{N} S_n(f) = \frac{1}{NP_w}\sum_{n=1}^{N}\left|\sum_{m=0}^{M_n-1} s^{(n)}(mT_n)e^{-j2\pi f mT_n}\right|^2. \tag{3.21}$$

This method is used for the purpose of reduction of shift of the received spectral assessments and reduction of the Gibbs effect (emergence of side lobes in a spectrum). It is especially expedient when temporal windows with low LSL are applied. Besides, overlapping of time intervals allows to increase a number of averaged periodograms with the set length of a pack (selection) of signals and by that to reduce the dispersion of assessment.

3.1.2 Parametric Methods of the Assessment of Moments of a Doppler Spectrum of the Signal Reflected by MO

The procedure of parametric spectral estimation includes three stages [3, 16]:

- the choice of a model of the temporal process (row);
  the estimation of parameters of the accepted model with use of the received readings of a signal on an observation interval (or the ACF values);
- the calculation of spectral assessments by means of substitution of the received model parameters in the calculated expressions for the power spectral density corresponding to this model.

3.1.2.1 Block Parametric Methods of Spectral Estimation

The block parametric methods as well as the nonparametric methods given above are intended for processing of the whole blocks of the accumulated readings of a signal of some fixed volume (packs). It is expedient to apply similar methods when the volume of the available pack is limited, but it is desirable to receive the assessment with the best possible characteristics. It is possible to consider the Yule-Walker, Levinson-Derbin, Berg methods, the covariant and the minimum norm methods, etc., as block parametric methods.

The Yule-Walker Method

The coefficients a_k of the AR of model of the reflected signal and its ACF are connected by the system of the linear equations [9, 17]. In particular, it is possible to write down the expression (2.106) for values of the temporal shift $q+1 \leq m \leq q+p$ and to present it in the matrix form

$$\begin{bmatrix} B(q) & B(q-1)\ldots & B(q-p) \\ B(q+1) & B(0) & \ldots B(q-p+1) \\ \ldots & \ldots & \ldots \\ B(q+p) & B(p-1)\ldots & B(q) \end{bmatrix} \begin{bmatrix} 1 \\ -a_1 \\ \ldots \\ -a_p \end{bmatrix} = \begin{bmatrix} B(q+1) \\ B(q+2) \\ \ldots \\ B(q+p+1) \end{bmatrix}. \tag{3.22}$$

Thus, if the ACF values are preset for $q+1 \leq m \leq q+p$, then the AR coefficients can be found as the solution of the system of the linear Eq. (3.22) called the Yule-Walker system of equations [2, 3, 18].

The matrix with the size $p \times p$ in the left part of the system (3.22) represents the ACM of the s(m) signal. If the s(m) signal is supposed by a stationary random process, then its ACM will be Toeplitz and Hermite [3].

For the AR model of the s(m) signal the AR coefficients a_k and the value σ_e^2 of the forming noise can be estimated by the assessments $\hat{B}(m)$ of values of unknown ACF of a signal. At the same time the expression (3.22) taking into account (2.105) is transformed to the form

$$\begin{bmatrix} B(0) & B(-1) & \dots & B(-p) \\ B(1) & B(0) & \dots & B(-p+1) \\ \dots & \dots & & \dots \\ B(p) & B(p-1) & \dots & B(0) \end{bmatrix} \begin{bmatrix} 1 \\ -a_1 \\ \dots \\ -a_p \end{bmatrix} = \begin{bmatrix} \sigma_e^2 \\ 0 \\ \dots \\ 0 \end{bmatrix}, \tag{3.23}$$

where $B(m) = \begin{cases} \frac{1}{M-m} \sum\limits_{i=0}^{M-m-1} s(i+m)s^*(i), & 0 \le m \le M-1, \\ \frac{1}{M-|m|} \sum\limits_{i=0}^{M-|m|-1} s(i+|m|)s^*(i), & -(M-1) \le m \le 0. \end{cases}$

The system of the Eq. (3.23) can be solved by means of the effective recursive algorithm developed independently by Levinson and Durbin [3] and demanding performance of the p^2 order computing operations.

The recurrent Levinson connects the AR coefficient a_k of the k order with the coefficient a_{k-1} of the $k-1$ order

$$a_{k,i} = a_{k-1,i} + \rho_k a_{k-1,k-i}^*, \tag{3.24}$$

where $i = 0,\dots,k$—the iteration number;
$\{a_{k,0} = 1,\ a_{0,i} = 1\}$—initial conditions;
$\rho_k = a_{k,k} = \frac{-\sum\limits_{i=0}^{k-1} a_{k-1,i} B(k-i)}{\sigma_{k-1}^2}$—the complex coefficients of reflection;
$\sigma_k^2 = \sigma_{k-1}^2 \left(1 - |\rho_k|^2\right)$—the assessment of the dispersion σ_e^2 of excitation noise, $\sigma_0^2 = B(0)$.

It should be noted that the MA coefficients are not linearly connected with the readings of the ACF and cannot be found by the solution of the systems of the linear equations similar to (3.25) or (3.23).

The iterative methods of optimization demanding big computational effort are used for the solution of the nonlinear equations at estimation of the MA coefficients. Besides they do not guarantee convergence or can even converge to incorrect solutions [3] therefore they, unlike the AR assessments are inexpedient to be applied to processing of signals in real time.

In case of packs of the reflected signal of large volume the direct solution of the Yule-Walker equations can give quite acceptable assessments of the DS [3] parameters, however when processing short packs the assessments received with its help have lower accuracy in comparison with other parametric methods.

Besides, the solution of the Yule-Walker system of the equations requires calculation of the ACF values. However, there are methods of determination of the AR model parameters based on direct use of readings of the input signal. These methods are based on close connection of the problem of identification of the AR parameters of process with theories of linear prediction and statistical estimation.

The Burg Method

The Burg algorithm, or the algorithm of the harmonious average [3, 18], is based on use of criterion of a minimum of RMS of an error of prediction, that is such values of the AR coefficients which for each value of the order of k minimize the sum of squares of errors of prediction forward (on readings $s(m-1)\ldots, s(m-\mathrm{k})$) and back (on readings of $s(m+1)\ldots, s(m+k)$) are found

$$\varepsilon_k = \sum_{m=k}^{M-1}\left[\left|e_k^f(m)\right|^2 + \left|e_k^b(m)\right|^2\right], \tag{3.25}$$

where $k = 1,\ldots, p$;

$$e_k^f(m) = s(m) - \hat{s}_f(m) = s(m) + \sum_{i=1}^{k} a_i^f s(m-i) \tag{3.26}$$

—forward prediction error;

$$e_k^b(m) = s(m-p) - \hat{s}_b(m-p) = s(m-p) + \sum_{i=1}^{k} a_i^b s(m-p+i) \tag{3.27}$$

—back prediction error.

The expressions (3.26) and (3.27) are in the form similar to the equation of the AR model (2.106), but with that difference that in case of processing of selections of readings of the finite size the errors $e_k^f(m)$ and $e_k^b(m)$ are not necessarily white noise therefore we will further assume that these errors are a bleached random process. It will allow to consider the AR coefficients a_i and the corresponding coefficients of forward a_i^f and back a_i^b linear prediction as equivalent.

Burg found that the value ε_k is the function of only one parameter—the complex coefficient of reflection ρ_k. Equating the complex derivative from ε on ρ_k to zero

$$\frac{\partial \varepsilon_k}{\partial \mathrm{Re}\{\rho_k\}} + j\frac{\partial \varepsilon_k}{\partial \mathrm{Im}\{\rho_k\}} = 0$$

and solving the received equation in relation to ρ_k, Burg received the following expression for the assessment of coefficients of reflection

$$\rho_k = \frac{-2\sum_{m=k}^{M-1} e_{k-1}^{f}(m+1)e_{k-1}^{b*}(m)}{\sum_{m=k}^{M-1}\left[\left|e_{k-1}^{f}(m+1)\right|^2 + \left|e_{k-1}^{b}(m)\right|^2\right]}. \quad (3.28)$$

The assessment (3.28) represents a harmonious average of coefficients of partial correlation of errors of forward and back prediction.

The Burg algorithm gives the shifted estimates of spectral components, and shift can reach 16% of resolution on frequency, characterized by the value 1/NT [3]. For reduction of this shift weighing of RMS error of prediction is used

$$\varepsilon_k = \sum_{m=k}^{M-1} w_k(m)\left[\left|e_k^{f}(m)\right|^2 + \left|e_k^{b}(m)\right|^2\right],$$

that leads to the following assessment of the coefficient of reflection

$$\rho_k = \frac{-2\sum_{m=k}^{M-1} w_{k-1}(m)e_{k-1}^{f}(m+1)e_{k-1}^{b*}(m)}{\sum_{m=k}^{M-1} w_{k-1}(m)\left[\left|e_{k-1}^{f}(m+1)\right|^2 + \left|e_{k-1}^{b}(m)\right|^2\right]}, \quad (3.29)$$

where $w_k(m)$—the weight function (for example, a square-law window or a Hamming window [19]).

Application of the considered parametric algorithms of spectral estimation (Yule-Walker, Burg) based on the use of the recursive procedure of Levinson-Derbin is complicated because of essential temporal not stationarity and spatial heterogeneity of the signals reflected from MO.

The Covariant Method of Linear Prediction

Determination of the AR model parameters by the Burg method is based on minimization of a RMS error of prediction in one parameter—the coefficient of reflectionρ_k. More general approach consists in joint minimization of RMS errors of prediction on all coefficients of forward and back linear prediction. The similar method called the covariant method consists in determination of such AR values of coefficients which for each value of the k order separately minimize the sum of squares of errors of forward and back prediction

$$\varepsilon_k^{f} = \sum_{m=k}^{M-1}\left[\left|e_k^{f}(m)\right|^2\right] \text{ and } \varepsilon_k^{b} = \sum_{m=0}^{k}\left[\left|e_k^{b}(m)\right|^2\right].$$

The expression for forward prediction of errors (3.26) of the p-th order in the range of temporal indexes from $m = 1$ to $m = M{+}p$ can be written down in the matrix form

$$\vec{E}_p^f = \underline{\mathbf{S}}_s \vec{a}^f,$$

where

$$\vec{E}_p^f = \left[e_p^f(0)\; e_p^f(1) \ldots e_p^f(p) \ldots e_p^f(M-p-1) \ldots e_p^f(M-1) \ldots e_p^f(M+p-1) \right]^T$$

$$\underline{\mathbf{S}}_s = \begin{bmatrix} s(0) & 0 & \ldots & 0 & 0 \\ s(1) & s(0) & \ldots & 0 & 0 \\ \ldots & \ldots & \ldots & \ldots & \ldots \\ s(p) & s(p-1) & \ldots & s(1) & s(0) \\ \ldots & \ldots & \ldots & \ldots & \ldots \\ s(M-p-1) & s(M-p-2) & \ldots & s(M-2p) & s(M-2p-1) \\ \ldots & \ldots & \ldots & \ldots & \ldots \\ s(M-1) & s(M-2) & \ldots & s(M-p-2) & s(M-p-1) \\ \ldots & \ldots & \ldots & \ldots & \ldots \\ 0 & 0 & \ldots & 0 & s(M-1) \end{bmatrix} = \begin{bmatrix} \underline{\mathbf{L}} \\ \underline{\mathbf{T}} \\ \underline{\mathbf{U}} \end{bmatrix}$$

—the rectangular $(M+p) \times (p+1)$ matrix of readings of the analysed signal; $\vec{a}^f = \left[1\; a_1^f \ldots a_p^f \right]^T$—$(p+1)$-element vector of coefficients of forward linear prediction;

$\underline{\mathbf{L}}$—lower triangular $p \times (p+1)$ matrix;
$\underline{\mathbf{T}}$—rectangular $(M-p) \times (p+1)$ matrix;
$\underline{\mathbf{U}}$—top triangular $p \times (p+1)$ matrix.

The use of matrixes $\underline{\mathbf{L}}$ and $\underline{\mathbf{U}}$ also assumes pre-processing and post-processing of the analysed signal by means of a rectangular weight window (that is equating to zero of the values before the first and after the last reading). In the analysis of only the available selection of readings without weighing it is necessary to consider $(M-p)$ the values of a signal in the expression for an error of prediction

$$\vec{e}_p^f = \left[e_p^f(p) \ldots e_p^f(M-1) \right]^T = \underline{\mathbf{T}} \vec{a}^f.$$

In this case the matrix equation minimizing the average square of an error of forward prediction

$$\varepsilon_k^f = \sum_{m=p}^{M-1} \left| e_p^f(m) \right|^2 = \left(\vec{e}_p^f \right)^H \vec{e}_p^f, \tag{3.30}$$

will have the form similar to the Yule-Walker Eq. (3.22)

$$\underline{\mathbf{T}}^H \underline{\mathbf{T}} \vec{a}^f = \vec{P}^f, \tag{3.31}$$

where $\vec{P}^f = \left[\rho_p^f \; 0 \ldots 0 \right]^T$—$(p+1)$-element vector.

The elements of Hermite $(p+1) \times (p+1)$-matrix $\underline{\mathbf{B}} = \underline{\mathbf{T}}^H \underline{\mathbf{T}}$ have the form of correlation forms [3]

$$B(i, j) = \sum_{m=p}^{M-1} s^*(m-i)s(m-j),$$

which cannot be written down as functions of a difference $(i-j)$, and it means that the autocorrelated matrix (ACM) $\underline{\mathbf{B}}$ is not a Toeplitz matrix [3, 20].

The necessary condition of not degeneracy of a matrix $\underline{\mathbf{B}}$ is the condition $(M-p) \geq p$ or $p \leq M/2$, that is the order p of the AR model should not exceed half of the length of selection of readings of the analysed signal.

Similar to the expressions (3.30) and (3.31) the matrix equation minimizing the average square of an error of back prediction

$$\varepsilon_k^b = \sum_{m=p}^{M-1} \left| e_p^b(m) \right|^2 = \left(\vec{e}_p^{\,b} \right)^H \vec{e}_p^{\,b}, \tag{3.32}$$

will have the form

$$\underline{\mathbf{T}}^H \underline{\mathbf{T}} \vec{a}^b = \vec{P}^b. \tag{3.33}$$

Here $\vec{e}_p^{\,b} = \left[e_p^b(p) \ldots e_p^b(M-1) \right]^T$—$(M-p)$-element vector of errors of back linear prediction; $\vec{a}^b = \left[a_p^b \ldots a_1^b \; 1 \right]^T$—$(p+1)$-element vector of coefficients of back linear prediction; $\vec{P}^b = \left[0 \ldots 0 \; \rho_p^b \right]^T$—$(p+1)$-element vector.

The solutions of the direct (3.31) and the inverse (3.33) matrix equations are interconnected as these equations in both cases contain the same ACM $\underline{\mathbf{B}}$. As both directions of prediction provide identical statistical information, it is expedient to unite statistics of errors of forward and back linear prediction to receive bigger number of points in which errors are defined (to increase the volume of the analysed selection of readings). Then $(M-p)$ of errors of forward and $(M-p)$ errors of back linear prediction can be united in one $2(M-p)$-element vector of error which has the following form

$$\vec{e}_p = \begin{bmatrix} \vec{e}_p^{\,f} \\ \vec{e}_p^{\,b*} \end{bmatrix} = \begin{bmatrix} \underline{\mathbf{T}} \\ \underline{\mathbf{T}}^* \underline{\mathbf{J}} \end{bmatrix} \vec{a},$$

where $\underline{\mathbf{T}} = \begin{bmatrix} s(p) & \ldots & s(0) \\ \ldots & \ldots & \ldots \\ s(M-p) & \ldots & s(M-2p) \\ \ldots & \ldots & \ldots \\ s(M-1) & \ldots & s(M-p-1) \end{bmatrix}$;

$\underline{\mathbf{J}}$—$(p+1) \times (p+1)$ matrix of reflection [20], such that

$$\underline{\mathbf{T}}^{*}\underline{\mathbf{J}} = \begin{bmatrix} s^{*}(0) & \ldots & s^{*}(p) \\ \ldots & \ldots & \ldots \\ s^{*}(M-2p) & \ldots & s^{*}(M-p) \\ \ldots & \ldots & \ldots \\ s^{*}(M-p-1) & \ldots & s^{*}(M-1) \end{bmatrix}.$$

Minimizing the average value of squares of errors of forward and back prediction,

$$\varepsilon_k = \frac{1}{2}\left[\sum_{m=p}^{M-1} \left|e_p^f(m)\right|^2 + \sum_{m=p}^{M-1} \left|e_p^b(m)\right|^2\right] = \frac{1}{2}\left(\vec{e}_p\right)^H \vec{e}_p = \frac{1}{2}\left[\left(\vec{e}_p^{\,f}\right)^H \vec{e}_p^{\,f} + \left(\vec{e}_p^{\,b}\right)^H \vec{e}_p^{\,b}\right], \tag{3.34}$$

it is possible to receive the following matrix equation [3]

$$\underline{\mathbf{B}}\vec{a} = \vec{P}, \tag{3.35}$$

in which the ACM elements

$$\underline{\mathbf{B}} = \begin{bmatrix} \underline{\mathbf{T}} \\ \underline{\mathbf{T}}^{*}\underline{\mathbf{J}} \end{bmatrix}^H \begin{bmatrix} \underline{\mathbf{T}} \\ \underline{\mathbf{T}}^{*}\underline{\mathbf{J}} \end{bmatrix} = \underline{\mathbf{T}}^H\underline{\mathbf{T}} + \underline{\mathbf{J}}\underline{\mathbf{T}}^T\underline{\mathbf{T}}^{*}\underline{\mathbf{J}}$$

have the form

$$B(i,j) = \sum_{m=p}^{M-1}\left[s^{*}(m-i)s(m-j) + s^{*}(m-p+i)s^{*}(m-p+j)\right]. \tag{3.36}$$

The procedure based on joint minimization of errors of forward and back linear prediction is called the modified covariant method [3, page 274]. Unlike the Burg method of based on minimization of only coefficient of reflection ρ_k (that is the prediction coefficient a_p), when using the modified covariant method the minimization is carried out on all coefficients of prediction.

The use of the modified covariant method is especially preferable in case of the temporal not stationarity and spatial heterogeneity of the s(m) signal reflected by MO leads to the fact that its ACM $\underline{\mathbf{B}}$ is symmetric (Hermite), but not Toeplitz [3, page 269]. Then application of the algorithm based on Cholesky [3, 21, 22] decomposition of

ACM $\underline{\mathbf{B}}$ on the lower and top triangular matrixes (triangularization by the method of a square root) is perspective [23, 24].

It is possible to reduce essentially the computational efforts at determination of unknown AR coefficients due to the property of hermiticity of the matrix $\underline{\mathbf{B}}$ (namely $B(i, j) = B(j, i)$) [24], thanks to which for definition of all its elements it is necessary to calculate only the lower triangular matrix from $\underline{\mathbf{B}}$.

Besides, the recursive formula

$$\begin{aligned} B(i+1, j+1) = B(i, j) + s^*(p-i)s(p-j) - s^*(M-i)s(M-i) \\ - s(i+1)s^*(j+1) + s(M-p+i+1)s^*(M-p+j+1). \end{aligned} \tag{3.37}$$

also reduces the number of necessary computing operations.

Thus, for finding of an autocorrelated matrix $\underline{\mathbf{B}}$ on a pack of readings of the s(m) signal it is enough to calculate one column by the formula (3.36), and to find other elements of the lower triangular matrix, using the recursive expression (3.37). The costs of definition of this matrix can be significantly reduced thanks to the fact that the matrix $\underline{\mathbf{B}}$ is the generalized Hermite matrix (that is $B(p-i, p-j) = B(j, i)$) [25].

The Eq. (3.35) has only one nonzero solution in relation to $\vec{a}$ provided that det $\underline{\mathbf{B}} \neq 0$, itdemonstrates positive definiteness of the matrix $\underline{\mathbf{B}}$. In this regard there is one and only one lower triangular matrix $\underline{\mathbf{L}}$ with positive diagonal elements, such that

$$\underline{\mathbf{L}\mathbf{L}}^H = \underline{\mathbf{B}}. \tag{3.38}$$

The Eq. (3.38) is the complex variant of Cholesky decomposition. It follows from (3.38) [21], that

$$\begin{cases} l(1,1) = \sqrt{B(1,1)}; & \\ l(1,j) = B(1,j)/\sqrt{B(1,1)}, & j > 1; \\ l(i,i) = \sqrt{B(i,i) - \sum\limits_{k=1}^{i-1} |l(k,i)|^2}, & i > 1; \\ l(i,j) = \frac{1}{l(i,i)} \sqrt{B(i,j) - \sum\limits_{k=1}^{i-1} l^*(k,i) l(k,j)}, & j > i. \end{cases} \tag{3.39}$$

The analysis of the ratios (3.39) shows what for the triangularization of the matrix $\underline{\mathbf{B}}$ of the size (*pxp*) it is necessary to make *p* extractions of a square root and about $p^3/6$ multiplications.

The matrix $\underline{\mathbf{L}}$ can be found by means of the recursive ratios [24]

$$\underline{\mathbf{B}}_p = \begin{bmatrix} \underline{\mathbf{B}}_{p-1} & \vec{r}_p \\ \vec{r}_p^H & B(p,p) \end{bmatrix}, \quad \underline{\mathbf{L}}_p = \begin{bmatrix} \underline{\mathbf{L}}_{p-1} & \vec{0} \\ \vec{l}_p & \lambda_p \end{bmatrix}, \tag{3.40}$$

where $\vec{r}_p = \left[\, B(p,0)\ B(p,1)\ \ldots\ B(p,p-1)\,\right]^H$;

$$\vec{l}_p = \vec{r}_p^{\,H} \underline{\mathbf{L}}_{p-1}^{-H}; \quad \lambda_p = \sqrt{B(p,p) - \vec{l}_p^{\,H}\vec{l}_p}.$$

For definition of the vector of the AR coefficients $\vec{a}$ we substitute decomposition (3.38) in the formula (3.35)

$$\underline{\mathbf{L}}_p \underline{\mathbf{L}}_p^H \vec{a} = \vec{P}.$$

Two equations correspond to this ratio

$$\underline{\mathbf{L}}_p^H \vec{a} = \vec{H}, \tag{3.41}$$

$$\underline{\mathbf{L}}_p \vec{H} = \vec{P}. \tag{3.42}$$

The recurrence ratio follows from the Eq. (3.42)

$$h_i = \frac{1}{l_p(i,i)}\left(P_i - \sum_{k=1}^{i-1} l_P^*(i,k) h_k\right), \quad i = \overline{1,p}. \tag{3.43}$$

The vector components of the AR coefficients are defined by the expression

$$a_i = \frac{1}{l_p(i,i)}\left(h_i - \sum_{k=i+1}^{p} l_P^*(k,i) a_k\right), \quad i = \overline{p,1}. \tag{3.44}$$

The Method of Minimum Dispersion

The main idea of the method of minimum dispersion developed by J. Capon for the solution of tasks of the space-time analysis of multidimensional arrays of sensors [3, 26], consists in finding of the vector $\vec{a}$ of the coefficients of the transversal filter which output signal [3, 27, 28]

$$y(n) = \sum_{k=0}^{p} a_k x(n-k) = \vec{a}^T \vec{x}(n),$$

at the set input signal $\vec{x}(n) = \left[\, x(n)\ x(n-1)\ \ldots\ x(n-p)\,\right]^T$ would have the maximum power.

Dispersion of the output signal of the filter is defined by the expression

$$\sigma_y^2 = \overline{|y(n)|^2} = \overline{\vec{a}^H \vec{x}^*(n) \vec{x}^T(n) \vec{a}} = \vec{a}^H \overline{\vec{x}^*(n) \vec{x}^T(n)} \vec{a} = \vec{a}^H \underline{\mathbf{B}} \vec{a},$$

where $\underline{\mathbf{B}}$—the known or estimated ACM of the input signal of the size (p + 1)×(p + 1).

Here it should be noted that the coefficients of the filter are chosen so that at an initial frequency of the analysis f_0 the filter frequency characteristic had single amplification factor, that is

$$\sum_{k=0}^{p} a_k \exp(-j2\pi f_0 kT) = \vec{a}^T \vec{e}^*(f_0) = \vec{e}^H(f_0)\vec{a} \equiv 1,$$

where $\vec{e}(f) = \left[1\ \exp(j2\pi fT)\ \ldots\ \exp(j2\pi fpT)\right]^T$—(p + 1)-the vector of complex sinusoids.

Then for finding of the vector $\vec{a}$ it is necessary to minimize the function of the type $\vec{a}^H\underline{\mathbf{B}}\vec{a} + \alpha\left(\vec{a}^H\vec{e}(\omega_0) - 1\right)$, where α—the Lagrange multiplier.

As a result the decision has the form [3, 26]

$$\vec{a} = \frac{\underline{\mathbf{B}}^{-1}\vec{e}(f_0)}{\vec{e}^H(f_0)\underline{\mathbf{B}}^{-1}\vec{e}(f_0)}, \tag{3.45}$$

and the power of an output signal of the forming filter at the frequency ω_0 is equal to

$$S(f_0) = T_n\sigma^2 = \frac{T_n}{\vec{e}^H(f_0)\underline{\mathbf{B}}^{-1}\vec{e}(f_0)},$$

where $-\frac{1}{2T_n} \le f_0 \le \frac{1}{2T_n}$.

Thus, the procedure of assessment of a spectrum by method of the minimum dispersion consists of the following steps:

1. The ACM $\underline{\mathbf{B}}$ is estimated on readings of the accepted signal.
2. Some initial (basic) frequency f_0 is set and, therefore, the basic vector $\vec{e}(f_0)$. $S(f_0)$ is calculated for it.

Then with some step the following value of basic frequency is introduced, and the calculations are repeated for it. The calculation of $S(f_0)$ is made in all range of possible frequencies in separate close enough located points. Then the position of maxima is found.

In connection with complexity of the implementation caused by need of the solution of the problem of search of a global extremum of multiextreme functionality for multidimensional space, the method of the minimum dispersion has limited practical application now.

The MUSIC Method

The method of multiple signal classification (MUSIC) [2, 3] is based on division of information which is contained in the ACM $\underline{\mathbf{B}}_x$ of the input mixture x (t) into two vector subspaces: signal subspace and noise subspace. These subspaces are defined by own vectors of ACM. In order that the vector $\vec{V}_i$ was the own vector of ACM, the fulfilment of the condition is necessary

$$\underline{\mathbf{B}}_x \vec{V}_i = \lambda_i \vec{V}_i,$$

where λ_i—the ACM own value corresponding to the vector $\vec{V}_i$.

As the rank of ACM $\underline{\mathbf{B}}_x$ is equal to M, this matrix will have M own vectors

$$\underline{\mathbf{B}}_x = \sum_{i=1}^{M} \vec{x}_i \vec{x}_i^H = \sum_{i=1}^{M} \lambda_i \vec{V}_i \vec{V}_i^H = \underline{\mathbf{V}}\,\underline{\Lambda}\,\underline{\mathbf{V}}^H,$$

where the own values λ_i of ACM are ordered on extent of decrease, that is

$$\lambda_1 \geq \lambda_2 \geq \ldots \geq^{\lambda_M},$$

and the own vectors $\vec{V}_i$ are orthogonal, that is

$$\vec{V}_i^H \vec{V}_j = \begin{cases} 1, & i = j; \\ 0, & i \neq j; \end{cases}$$

where $\underline{\mathbf{V}} = \left[\, \vec{V}_1 \; \vec{V}_2 \, \ldots \, \vec{V}_M \,\right]$—the matrix of the MhM size which columns are own vectors of ACM;
$\underline{\Lambda} = \mathrm{diag}\left[\, \lambda_1 \; \lambda_2 \, \ldots \, \lambda_M \,\right]$—the diagonal matrix.

If the input mixture consists of the useful signal representing readings of narrow-band process (2.32), and white noise, then its ACM can be presented as the sum of ACM of a signal $\underline{\mathbf{B}}$ and noise $\underline{\mathbf{B}}_n$

$$\underline{\mathbf{B}}_x = \sum_{i=1}^{M} \vec{x}_i \vec{x}_i^H = \sum_{i=1}^{M} (\vec{s}_i + \vec{n}_i)(\vec{s}_i + \vec{n}_i)^H = \sum_{i=1}^{M} \vec{s}_i \vec{s}_i^H + \sum_{i=1}^{M} \vec{n}_i \vec{n}_i^H = \underline{\mathbf{B}} + \underline{\mathbf{B}}_n, \tag{3.46}$$

where $\underline{\mathbf{B}} = \sum_{i=1}^{M} \vec{s}_i \vec{s}_i^H$—ACM of a useful signal;
$\underline{\mathbf{B}}_n = \sum_{i=1}^{M} \vec{n}_i \vec{n}_i^H = \rho_n \underline{\mathbf{I}}$—ACM of intra reception noise;
ρ_n—noise dispersion;
$\underline{\mathbf{I}}$—single M × M matrix.

If the AR model order p is less than the rank of the ACM of input mixture determined by the value M of the processed pack, then decomposition of ACM on own values (3.46) can be brought to the following form

$$\underline{\mathbf{B}}_x = \sum_{i=1}^{M} \lambda_i \vec{V}_i \vec{V}_i^H = \sum_{i=1}^{p} \lambda_i \vec{V}_i \vec{V}_i^H + \rho_n \sum_{i=1}^{M} \vec{V}_i \vec{V}_i^H$$
$$= \sum_{i=1}^{p} (\lambda_i + \rho_n) \vec{V}_i \vec{V}_i^H + \sum_{i=p+1}^{M} \rho_n \vec{V}_i \vec{V}_i^H. \tag{3.47}$$

From (3.47) it follows that p of the greatest (so-called main) own vectors $\vec{V}_1 \ldots \vec{V}_p$ will correspond to a useful signal (signal subspace). The remained (M − p) vectors will correspond to clean noise (noise subspace). Theoretically, as it follows from (3.47), all (M − p) of the smallest own values have to coincide in value, however really because of existence of errors of measurements they are just very close [2, 3].

As own vectors of ACM are mutually orthogonal and the main own vectors are in the same subspace, as signal vectors, the signal vectors are orthogonal to all vectors in noise subspace, including any their combination,

$$\vec{s}_i^H \left[\sum_{k=M+1}^{p} \alpha_k \vec{V}_k \right] = 0 \text{ or } \sum_{k=M+1}^{p} \alpha_k \left| \vec{s}_i^H \vec{V}_k \right|^2 = 0, \tag{3.48}$$

where α_k—the weight coefficient.

Assuming all $\alpha_k = 1$, we transform the Eq. (3.48) to the form

$$\sum_{k=M+1}^{p} \left| \vec{s}_i^H \vec{V}_k \right|^2 = \sum_{k=M+1}^{p} \vec{s}_i^H \vec{V}_k \vec{V}_i^H \vec{s}_i = \vec{s}_i^H \left(\sum_{k=M+1}^{p} \vec{V}_k \vec{V}_i^H \right) \vec{s}_i \equiv 0,$$

where the expression

$$D(f) = \vec{s}_i^H \left(\sum_{k=M+1}^{p} \vec{V}_k \vec{V}_i^H \right) \vec{s}_i$$

has the name "zero range".

The signal spectrum is defined by value, inverse to "zero range", [27]

$$S(f) = \frac{1}{D(f)} = \frac{1}{\vec{s}_i^H \left(\sum_{k=M+1}^{p} \vec{V}_k \vec{V}_i^H \right) \vec{s}_i}. \tag{3.49}$$

The assessment (3.49) is not the valid assessment of the spectrum of the accepted signal as it is impossible to determine the absolute value of the power of spectral components by it, and in some situations it is impossible to determine relative values, that is there is no correct inverse transformation [29]. The expression (3.49) represents the spectrum pseudo-assessment by which it is possible to determine with high accuracy only a set of frequency components of an initial signal.

A number of the parametric methods, similar to MUSIC, based on the analysis of the ACM own values of signals also has similar properties:

- the minimum norm method [27, 30] consisting in the assessment of spectral components of a signal on the basis of the ratio received by algebraic transformations

$$S(f) = \frac{\left(\vec{u}_1^H \underline{\mathbf{V}}_n \underline{\mathbf{V}}_n^H \vec{u}_1\right)^2}{\left|\vec{e}^H(f)\underline{\mathbf{V}n}\,\underline{\mathbf{V}}_n^H \vec{u}_1\right|^2}, \tag{3.50}$$

where $\vec{u}_1$—of the size $(p + 1)$, such that $u_{1i} = \begin{cases} 1, \; i = 1 \\ 0, \; i > 1 \end{cases}$,

$\underline{\mathbf{V}}_n = \left| \vec{V}_{M+1} \; \ldots \; \vec{V}_{p+1} \right|$—the matrix of the size (p + 1) × (p − M), made of noise own vectors of the ACM $\underline{\mathbf{B}}$.

Thus, for search of the spectrum it is necessary to estimate previously the matrix $\underline{\mathbf{V}}_n$ of noise own vectors, and the expression in the numerator (3.50) represents the constant not depending on *f*.

The ESPRIT (Estimation of Signal Parameter via Rotational Invariance Technique) method [3, 31] consists in frequency assessment in a signal subspace and is based, unlike MUSIC, on use of the main own vectors $\vec{V}_1 \ldots \vec{V}_p$ (3.47) of the signal ACM. The procedure of assessment of signal frequency by this method is based on replacement of ACM with its approximation of the lowered rank which is written down through the main own vectors.

3.1.2.2 The Recursive Parametric Methods of Spectral Estimation

For the solution of the problem of the assessment of the DS parameters of the spatially non-uniform and non-stationary signals reflected by MO the adaptive recursive (consecutive) procedures of estimation consisting in step-by-step correction of the current assessments of the moments (1.16)–(1.18) of the spectrum [32, 33] can be used. The recursion proceeds until errors of the assessment of the parameters of the MO spectrum become less than the set value. The flowing methods belong to the recursive:

1. The gradient methods (the method of the fastest descent, the method of minimization of a root mean square error (RMSE), differential algorithms of the fastest descent, the accelerated gradient algorithms and other methods differing in procedures of the assessment of a gradient of an error and the amount of a step of search [34]). Their common fault is the high sensitivity to the dispersion of the ACM own values worsening convergence of recursive assessment procedure.

Let us note that the dispersion of own values can be estimated by means of a number of conditionality of the ACM cond $\underline{\mathbf{B}} = \lambda_{\max}/\lambda_{\min}$. At R sounding of MO by means of an airborne pulse-Doppler radar station the number of conditionality of the ACM of an input signal, equal to SNR, can reach several tens decibels [20].

2. The Least Square Methods (MSS) Based on Step-by-Step Minimization of RMS Error of Linear Prediction. In the classical variant of MSS the error $\varepsilon_p = (\vec{e}_p)^H \vec{e}_p$, where $\vec{e}_p$—the vector of errors entering the matrix Eq. (3.40)

$$\underline{\mathbf{B}}\vec{a} + \vec{e}_p = \vec{P}. \tag{3.51}$$

Equating the local derivative $\partial\varepsilon/\partial\vec{a}$ to zero (on the necessary condition of existence of an extremum), the vector assessment to be found is $\vec{a}$

$$\vec{a} = \left[\frac{\underline{\mathbf{B}}^H \vec{P}}{\underline{\mathbf{B}}^H \underline{\mathbf{B}}}\right]^H. \tag{3.52}$$

Having substituted (3.52) in (3.51), we receive the MSS assessment of the vector of the AR coefficients

$$\overset{\wedge}{\vec{a}} = (\underline{\mathbf{B}}^H \underline{\mathbf{B}})^{-1} \underline{\mathbf{B}}^H \vec{P} = (\underline{\mathbf{B}}^H \underline{\mathbf{B}})^{-1} \underline{\mathbf{B}}^H (\underline{\mathbf{B}}\vec{a} + \vec{e}_p) = \vec{a} + (\underline{\mathbf{B}}^H \underline{\mathbf{B}})^{-1} \underline{\mathbf{B}}^H \vec{e}_p. \tag{3.53}$$

The second summand in (3.53) is equal to zero only if the vector of random errors $\vec{e}_p$ is equal to zero, that is the assessment $\overset{\wedge}{\vec{a}}$ coincides with exact value of the vector $\vec{a}$ only at the total absence of random errors.

In classical variant of LSM a priori dispersions of each separate reading are accepted as equal to unit and do not enter evidently in the expression (3.53). Assuming inequality of the received readings of signals, it is necessary to introduce the corresponding matrix of covariants $\underline{\mathbf{Q}}_\varepsilon = E\{\vec{\varepsilon}_p \vec{\varepsilon}_p^H\}$, which enters the vector assessment equation $\vec{a}$ for the weighed LSM (weighted least squares method) on the basis of the priori assumptions of the accuracy of each separate reading

$$\overset{\wedge}{\vec{a}} = (\underline{\mathbf{B}}^H \underline{\mathbf{Q}}_\varepsilon^{-1} \underline{\mathbf{B}})^{-1} \underline{\mathbf{B}}^H \underline{\mathbf{Q}}_\varepsilon^{-1}. \tag{3.54}$$

If assessment errors on different iterations are correlated, then the matrix of a priori covariants $\underline{\mathbf{Q}}_\varepsilon$ is full. This version of the LSM is called the generalized LSM (generalized least squares method). Because it is, as a rule, difficult to calculate aprioristic correlations between observations, the generalized LSM in pure form is practically not applied.

3. The Recursive Method of Assessment on the Basis of Kalman Filter.

In terms of optimum linear discrete filtration the problem of assessment of the AR coefficients is as follows [35]. Let's assume that the readings of the random sequence *x*(*m*) which is an additive mixture of the *s*(*m*) signal and *n*(*m*) broadband noise are available to observation at the input of the block of assessment

$$x(m) = s(m) + n(m). \tag{3.55}$$

The useful *s*(*m*) signal is the function of time t and the multicomponent message $\vec{u} = \left[a_1\ a_2\ \ldots\ a_p\right]^T$, representing a vector random process,

$$s(m) = \sum_{k=1}^{p} a_k(m)s(m-k) + e(m) = \vec{a}^T(m)\vec{s}(m) + e(m), \tag{3.56}$$

where $\vec{s}(m) = \left[s(m-1)\ s(m-2)\ \ldots\ s(m-p)\right]^T$—the vector of readings of an output signal of the forming filter;
$\vec{a}(m) = \left[a_1(m)\, a_2(m) \ldots a_p(m)\right]$—the message vector assessment $\vec{a}$ at the m-th step.

The noises n[m] and e[m] are assumed as the normal white independent random sequences, such that conditions are satisfied

$$\overline{n(m)n^*(i)} = \sigma_n^2\delta(m-i);\ \overline{e(m)e^*(i)} = \sigma_e^2\delta(m-i);\ \overline{n(m)e^*(i)} = 0,$$

where $\delta(m-i) = \begin{cases} 1, & m = i; \\ 0, & m \neq i \end{cases}$ —the Kronecker delta;
σ_n^2, σ_e^2—the dispersions of the sequences *n*[*m*] and *e*[*m*] respectively.

In this case the problem of filtration consists in that on realization of the process *x*(*m*) to estimate optimally the values of components of the vector $\vec{a}$.

All available information on the current values of components of the vector $\vec{a}$, received in the course of observation of process *x*(*m*) contains in the posteriori conditional density of probabilities $p\left[\vec{a}(m) : m|x(1) \ldots x(m)\right]$ of the vector $\vec{a}$ [36].

Knowing the posteriori density of probabilities $p\left[\vec{a}(m) : m|x(1) \ldots x(m)\right]$, it is possible to create the current assessment of the vector $\vec{a}$, for which the vector corresponding to the maximum $p\left[\vec{a}(m) : m|x(1) \ldots x(m)\right]$ is taken.

The posteriori density $p\left[\vec{a}(m) : m|x(1) \ldots x(m)\right]$ changes both under the influence of the changing process $\vec{a}$, and as a result of accumulation of data on process $\vec{a}$ on the accepted process *x*(*m*). Change of the process $\vec{a}$ leads to widening of $p\left[\vec{a}(m) : m|x(1) \ldots x(m)\right]$, and accumulation of information about it—to narrowing of this distribution.

On the basis of the prior data given above the assessment, optimum by the criterion of maximum posteriori probability $\vec{a}$ is defined by the following recurrent equations [35, 36]:

(a) the observation equation

$$x(m) = \vec{a}^T(m)\vec{s}(m) + e(m) + n(m); \tag{3.57}$$

(b) the equation of optimum assessment

$$\hat{\vec{a}}(m) = \hat{\vec{a}}(m-1) + \vec{k}(m)\left[x(m) - \hat{\vec{a}}^T(m-1)\vec{s}(m)\right]; \tag{3.58}$$

(c) the equation of optimum weight coefficients

$$\vec{k}(m) = \frac{\gamma(m-1)\vec{s}^*(m)}{\sigma_\Sigma^2 + \gamma(m-1)\vec{s}^T(m)\vec{s}^*(m)} = \frac{\gamma(m-1)\vec{s}^*(m)}{\sigma_n^2 + \sigma_e^2 + \gamma(m-1)|\vec{s}(m)|_E^2}; \tag{3.59}$$

(d) the assessment error equation

$$\gamma(m) = \left[1 - \vec{k}(m)\vec{s}^T(m)\right]\gamma(m-1) = \gamma(m-1) - \frac{\gamma(m-1)^2|\vec{s}(m)|_E^2}{\sigma_n^2 + \sigma_e^2 + \gamma(m-1)|\vec{s}(m)|_E^2}, \tag{3.60}$$

where $|\vec{s}(m)|_E^2 = \vec{s}^T(m)\vec{s}^*(m)$—the square of Euclidean norm of the vector $\vec{s}(m)$.

The Eqs. (3.57)–(3.60) completely describe the algorithm of the optimum linear Kalman filter (the Fig. 3.1). At an input of the Kalman filter the assessment of the accepted additive mixture x(m) received on the previous observations (a predictable part), $\hat{\vec{a}}^T(m-1)\vec{s}(m)$ is deducted from it. From the linear combination of this difference weighed with the weight $\vec{k}(m)$, and from the previous assessment (the prior data) $\hat{\vec{a}}(m-1)$ the current optimum assessment is formed $\hat{\vec{a}}(m)$.

At the non-stationary nature of reflections from MO the nonunitary transitional matrix of a state reflecting interrelation in time of consecutive values of the vector $\vec{a}$ should be introduced into the Eqs. (3.57)–(3.60) [35].

The initial conditions in the equations can be set in the form is known.

$$\hat{\vec{a}}(0) = \left[0\ 0\ \ldots\ 0\right], \quad \gamma(0) = 1 \tag{3.61}$$

Let's assume now that the vector $\vec{a}$ is known. Then for the signal and interference situation described by the Eq. (3.55), the optimum recurrent extrapolation of the s(m) signal is defined by the differential equations which are a special case of the equations of the Kalman-Biussi filter [35, 36]

$$\hat{s}(m+1) = \vec{a}^T\vec{s}(m) + \frac{\left[\sigma_e^2 + p(m)|\vec{a}|_E^2\right]\left[x(m+1) - \vec{a}^T\vec{s}(m)\right]}{\sigma_n^2 + \sigma_e^2 + p(m)|\vec{a}|_E^2}, \tag{3.62}$$

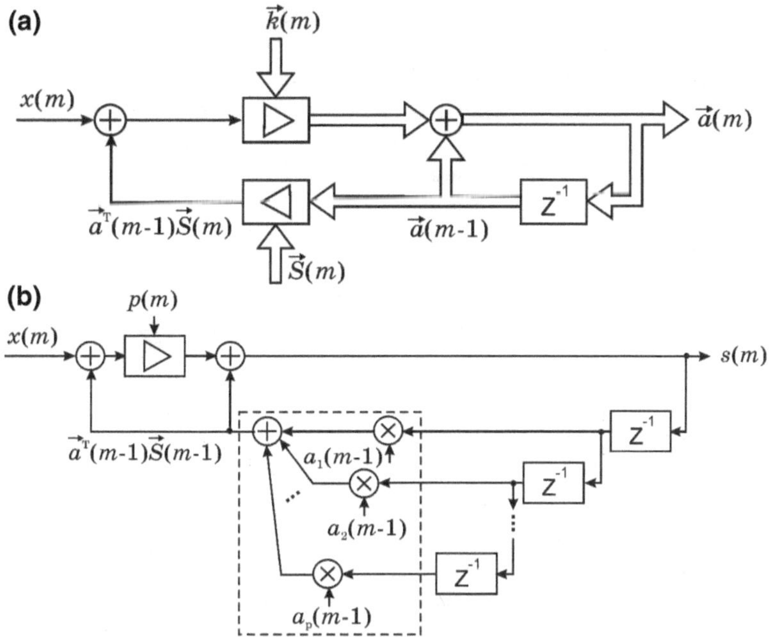

Fig. 3.1 The block diagram of the device of recursive assessment: **a** the block of assessment of the AR coefficients; **b** the block of filtration of a useful signal

$$\text{where} \quad p(m+1) = \left[\sigma_e^2 + p(m)|\vec{a}|_E^2\right] - \frac{\left[\sigma_e^2 + p(m)|\vec{a}|_E^2\right]^2}{\sigma_n^2 + \sigma_e^2 + p(m)|\vec{a}|_E^2} \tag{3.63}$$

– the current value of a square of a root mean square error of filtration.

The Eqs. (3.62) and (3.63) describe the linear discrete filter having p complex poles and changing in time transfer coefficient.

Using the mutual substitution of the current solutions of the Eqs. (3.57)–(3.60) and (3.62), (3.63), we receive the system of the equations describing the operation of the adaptive discrete filter and allowing to carry out the recurrent assessment of the vector $\vec{a}$ of the AR coefficients in process of receipt of the reflected signals. After accumulation the average frequency $\bar{f}$ and width Δf of DS of the signal reflected by MO are determined by the assessment of the AR coefficients.

Let us note that at the solution of the specific applied objectives the application of the Kalman filtration is accompanied by a number of such problems as the possible accumulation of errors, the probable absence of convergence of process of estimation arising because of incompleteness or insufficiency of prior information on properties of signals or possible loss of operability of the filter when feeding to its input of the signal which does not have noise in any of the components [37].

The advantage of the recursive algorithms is the ability of correction of intermediate errors in the course of assessment that is important at strong temporal not stationarity and spatial heterogeneity of the processed signals. The main disadvantage of the majority of recursive methods is the large volume of the calculations which are carried out on each step of a recursion in real time and also the low velocity of convergence of the received assessments to the true values of the parameters.

3.1.3 The Comparative Analysis of the Methods of Estimation of Parameters of a Doppler Spectrum of the Signal Reflected by MO

Nonparametric methods of the spectral analysis, owing to limitation of frequency resolution and accuracy of determination of frequencies of separate DS components of the reflected signals on selections of limited length, do not provide performance of technical characteristics on the accuracy and reliability of detection of dangerous zones of WS and turbulence in MO (the Sect. 1.3). The numerous ways of improvement of classical methods of spectral assessments offered recently (for example, the application of temporal or spectral windows or addition of selection by zeros) allow to improve somehow the accuracy of assessment of separate spectrum components, but do not allow to increase the frequency resolution.

The analysis of the possibility of use for the solution of the problem of assessment of the moments of DS of MO signals of parametric methods is of interest.

The methods based on the AR process statistical model are the most preferable for the analysis of DS of the signals reflected by MO. It is connected, first of all, with their ability to allocate spectrum maxima at separate frequencies. Other models on the basis of rational functions (MA and ARMA) do not have this advantage or are now insufficiently investigated.

Application of the traditional parametric methods of Yule-Walker and Burg based on use of the recursive procedure of Levinson-Derbin for the assessment of the DS parameters, is complicated because of essential heterogeneity and not stationarity of the signals reflected by MO. Besides, the Burg method is less effective in comparison with other parametric methods (covariant, the minimum dispersion and MUSIC) in the presence within the allowed volume of several local groups of HM moving with different velocities [38].

Another important disadvantage of the Burg method of is breaking of spectral lines of narrow-band process in its analysis against the background of broadband noise [3]. Conditions under which such breaking takes place (high SNR, the big relative order of the model equal to the model order ratio to the length of selection of the analysed signal) often arise when sounding MO by the airborne RS installed on mobile carriers.

The use of the covariant method having the largest accuracy among the AR methods of assessment of spectral components [3] is more preferable. The methods based on the analysis of the ACM own values of the accepted signals (MUSIC, ESPRIT [31, 39], the minimum norm, etc.) have comparable characteristics on the accuracy of frequency assessments, and though the MUSIC method is poorly comparable to the minimum norm method in the accuracy of estimation [40], but it is steadier against instability of the amplification factor of a reception path and presence of phase errors [41]. The ESPRIT method is intermediate. However, the focus of the specified methods on allocation of signals of the point targets against the background of broadband noise complicates their use at the solution of the problem of the assessment of the moments of DS of such spatially distributed target as MO. In particular, in [42] it is noted that the methods based on the analysis of the ACM own values lead to wrong (ambiguous) measurements in all cases when MO which are the components of DS of signals are strongly correlated. The sufficient condition of it is the sparseness of the selection of readings of signals [43].

The preliminary conclusion that the most preferable method for realization in airborne RS is the modified covariant method and MUSIC method follows from the submitted analysis of properties of various parametric methods of the assessment of the moments of DS of the reflected signal. For the reasonable choice it is necessary to compare the considered methods by a number of criteria: the adequacy and accuracy of the assessment, computing complexity, stability and sensitivity to stability of amplitude and phase characteristics of channels of a reception path of airborne RS, the level of own noise, the correlation and level of the accepted signals, the volume of a pack of signals used for the assessment of the spectral moments.

Comparison of characteristics of the assessments of the DS parameters provided with the chosen parametric methods with the results achieved when using classical methods of spectral estimation is of also interest.

3.1.3.1 The Analysis of Accuracy of the Assessment of the Moments of a Doppler Spectrum

The total error (error) of measurement of the moments of DS of the signals reflected by the allowed MO volume contains systematic, dynamic and fluctuation components. The change of the most measured value during its assessment is the reason of a dynamic error. The systematic error of measurement is caused by an error of setup of filters in paths of formation of PS and processing of the accepted signals. The sources of a fluctuation error are internal noise of the RS receiver, the external interferences, the fluctuations of the most useful signal. Let's consider further the fluctuation error of the assessment of the moments of DS of the reflected signals characterized by the dispersion of the corresponding assessments [44].

The highest accuracy (the minimum dispersion) of the assessment at the large volume M of a pack of signals is provided by the ML method (the Sect. 3.1.1.1) [45]. At the same time as the assessment of the measured parameter ε of the reflected signal its value corresponding to the maximum of likelihood function used. The dispersion of assessment of the parameter ε, determined by the ratio [6]

$$\sigma_{\varepsilon}^{2} \geq -\frac{1}{\left\langle \partial^{2}/\partial\varepsilon^{2} \ln L(x[t]\,|\varepsilon)\right\rangle}, \tag{3.64}$$

sets the lower bound, the so-called Cramer-Rao bound [35, 46] with which the dispersions of assessments of other algorithms should be compared. Here $L(x[t]\,|\varepsilon)$—the likelihood function. In case DS of an MO signal has the Gaussian form, and intra reception noise is an uncorrelated normal random process, then the dispersion of assessment of average frequency $\bar{f}$ by the ML method can be presented in the form [9, 47]

$$\sigma_{\bar{f}}^{2} \geq \frac{12\Delta f^{4} T_{n}^{2}}{M\left[1 - 12\left(\Delta f T_{n}\right)^{2}\right]} \tag{3.65}$$

at big SNR and the expression

$$\sigma_{\bar{f}}^{2} \geq \frac{2}{\sqrt{\pi}} \frac{\Delta f^{3} T_{n}}{M} Q^{-2} \tag{3.66}$$

at small SNR

The dispersion of the assessment of the width Δf of DS are presented by the similar expressions [9, 47]

$$\sigma_{\Delta f}^{2} \geq \frac{45}{4} \frac{\Delta f^{6} T_{n}^{4}}{M} \tag{3.67}$$

at big SNR and

$$\sigma_{\Delta f}^{2} \geq \frac{2}{\sqrt{\pi}} \frac{\Delta f^{3} T_{n}}{M} Q^{-2} \tag{3.68}$$

at small SNR

For the assessment of expediency of realization of parametric methods of spectral estimation it is necessary to compare their accuracy characteristics to the corresponding expressions for classical nonparametric methods therefore we will give the ratios determining the accuracy of assessment of $\bar{f}$ and Δf by the methods of paired pulses and periodograms below.

The dispersions of assessments of the average frequency $\bar{f}$ and width Δf of DS of the signals reflected by MO, by the method of paired pulses, received on the basis of the variation analysis [48] are equal respectively [49] to

$$\sigma_{\bar{f}}^{2} = \frac{\left(1 + Q^{-1}\right)^{2} - \rho^{2}(T_{n})}{8\pi^{2} M \rho^{2}(T_{n}) T_{n}^{2}} \tag{3.69}$$

$$\sigma^2_{\Delta f} = \begin{cases} \frac{3}{8\sqrt{\pi}} \frac{\Delta f}{MT_n} & \text{at big SNR} \\ \frac{1,5}{\Delta f^2 T_n^4 MQ} & \text{at small SNR} \end{cases}, \qquad (3.70)$$

where $\rho(T_n) = \exp\left[-(\pi \Delta f T_n)^2\right]$—the rated correlation coefficient.

From (3.69) it follows that the dispersion of the assessment $\bar{f}$ exponential increases at increase of Δf or T_p.

For the periodogram method the expressions defining dispersions of the assessments of the average frequency $\bar{f}$ and width Δf of DS of signals of MO respectively have the form [7, 9]

$$\sigma^2_{\bar{f}} = \frac{1}{MT_n^2}\left[\frac{\sqrt{\pi}\Delta f T_n}{2} + 8(\pi \Delta f T_n)^2 Q^{-1} + \frac{1}{12}Q^{-2}\right], \qquad (3.71)$$

$$\sigma^2_{\Delta f} = \frac{1}{MT_n^2}\left[\frac{3\Delta f T_n}{8\sqrt{\pi}} + 4(\pi \Delta f T_n)^2 Q^{-1} + \left(\frac{1}{320\Delta f^2 T_n^2} + \frac{\Delta f^2 T_n^2}{4} - \frac{1}{24}\right)Q^{-2}\right]. \qquad (3.72)$$

Let us notice that the first member (3.71) is equal to the corresponding member in the expression received at expansion in a series of the formula (3.69) for the method of paired pulses. It means that the efficiency of both methods at high SNR is comparable, however other members (3.71) exceed the corresponding members (3.69). Therefore, at the small values of SNR and the narrow spectrum ($\Delta f T_p < 1/4$), the method of paired pulses gives the assessments of $\bar{f}$ with smaller dispersion.

In the expression (3.72) for the dispersion of assessment of spectrum width by the periodogram method the first member is also equal to the value (3.70) of the dispersion of assessment by the method of paired pulses at high SNR. It means that at big SNR these methods are equivalent. At low SNR the dispersion $\sigma^2_{\Delta f}$ of the assessment of the spectrum width of by the method of paired pulses will be approximately three times less, than the value received by the periodogram method.

Here it should be noted that increase in dispersions of the assessments of $\bar{f}$ and Δf by the periodogram method with the growth Δf is not the subject to the exponential law that does the use of this method preferable in case of wide spectra.

The dependence of dispersion of the assessment of the width of DS of the signals reflected by MO from the selection length (pack volume) is presented in Fig. 3.2a. When calculating it was supposed that SNR Q = 10 dB, $T_p = 1$ ms, $\sigma_V = 5$ m/s (that corresponds to heavy turbulence), λ = 3 cm.

The given results confirm the inversely proportional dependence between $\sigma^2_{\Delta f}$ and M, following from (3.67), (3.70) and (3.72). The same dependence, but in other scale—for the RMSE of the assessment of pulsations of radial velocity for the allowed MO volume is presented in Fig. 3.2b. As an airborne RS should provide the accuracy of assessment of the width of the spectrum of radial velocity of MO of the order of ±1 m/s, then proceeding from the Fig. 3.2, the necessary volume of a pack will be from 7 (the ML method) to 24 (the periodogram method) the reflected impulses.

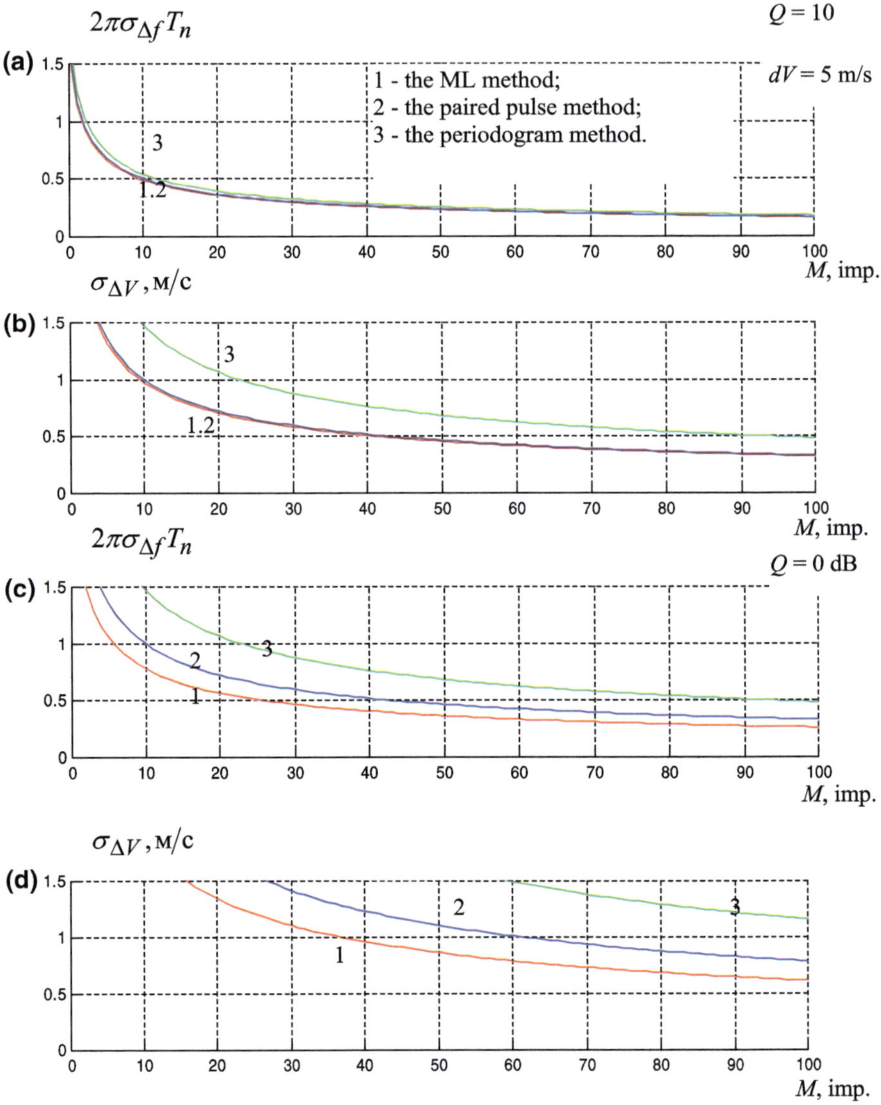

Fig. 3.2 Dependence of RMS errors of assessment of the rated width of DS of an MO signal (**a**, **c**) and pulsations of radial velocity (**b**, **d**) from the selection length (pack volume) for the various values of SNR

The dependences of the dispersion of assessment of the width of a DS of the signals reflected by MO and the RMS of an error of the assessment of the width of the spectrum of radial velocity of MO from a pack volume received at smaller SNR (Q ==0 dB) are given in Fig. 3.2c and d. The decrease of SNR can be caused, for example, by presence of broadband uncorrelated passive interferences. In this case the minimum admissible volume of a pack is from 40 (the ML method) to 120 (the periodogram method) impulses.

In Fig. 3.3 the dependences of the dispersion of assessment of average frequency of DS of the reflected signals rated to T_p, and the dispersion of assessment of the average radial velocity of the allowed MO volume from the selection length (pack volume) received at the same basic data are presented. For provision of the assessment of average radial speed with the accuracy not more than ± 1 m/s the necessary volume of a pack is from 8 (the ML method, SNR Q = 10 dB) to 200 (the periodogram method, SNR Q = 0 dB) reflected impulses.

The comparison of the results given in Figs. 3.2 and 3.3 show that reduction of SNR influences the accuracy of the assessment of the average radial velocity much stronger (that is the accuracy of the assessment of degree of WS danger) in comparison with the accuracy of assessment of the RMS width of the spectrum of radial velocities characterizing atmospheric turbulence.

In Figs. 3.4 and 3.5 the dependences of the dispersion of the assessment of the width and average frequency of DS of the reflected signals and also the average and RMS value of the radial velocity of particles of the allowed MO volume, on intensity of the atmospheric turbulence determined by the RSME value of the radial velocity are presented. These dependences confirm the earlier made conclusion that at reception of rather narrow-band signals reflected by MO the use of the method of paired pulses is rather effective [50]. Rather small width of DS of the reflected signal takes place at weak and moderate atmospheric turbulence, small velocities of flight of the RS carrier and narrow AS DP. At widening of the spectrum of the reflected signal the accuracy of the assessment by this method decreases exponentially. The advantages of the periodogram method are shown at reception of the reflected signals with the wide DS caused by turbulence, movement of the RS carrier and scanning of AS DP.

Reduction of the SNR value determined, first of all, by the specific MO EER leads to significant deterioration in the accuracy of the assessments of DS parameters. Partially it is possible to reduce the impact of decrease in the specific MO EER (for example, in case of cloud cover with small density) by means of coherent accumulation of the signals reflected by the allowed MO volume provided with significant increase in selection (that is due to increase in time of processing of the reflected signals).

Let us carry out the analysis of accuracy of the assessment of the moments of DS of the signals reflected by MO by parametric methods. As it is established above, in this case the most preferable are the modified covariant method and the MUSIC method. The RMS error values of the assessments received with use of imitating mathematical modelling for which algorithm flowchart is given in Figs. 3.6 and 3.7 are used as the parameter characterizing the assessment accuracy.

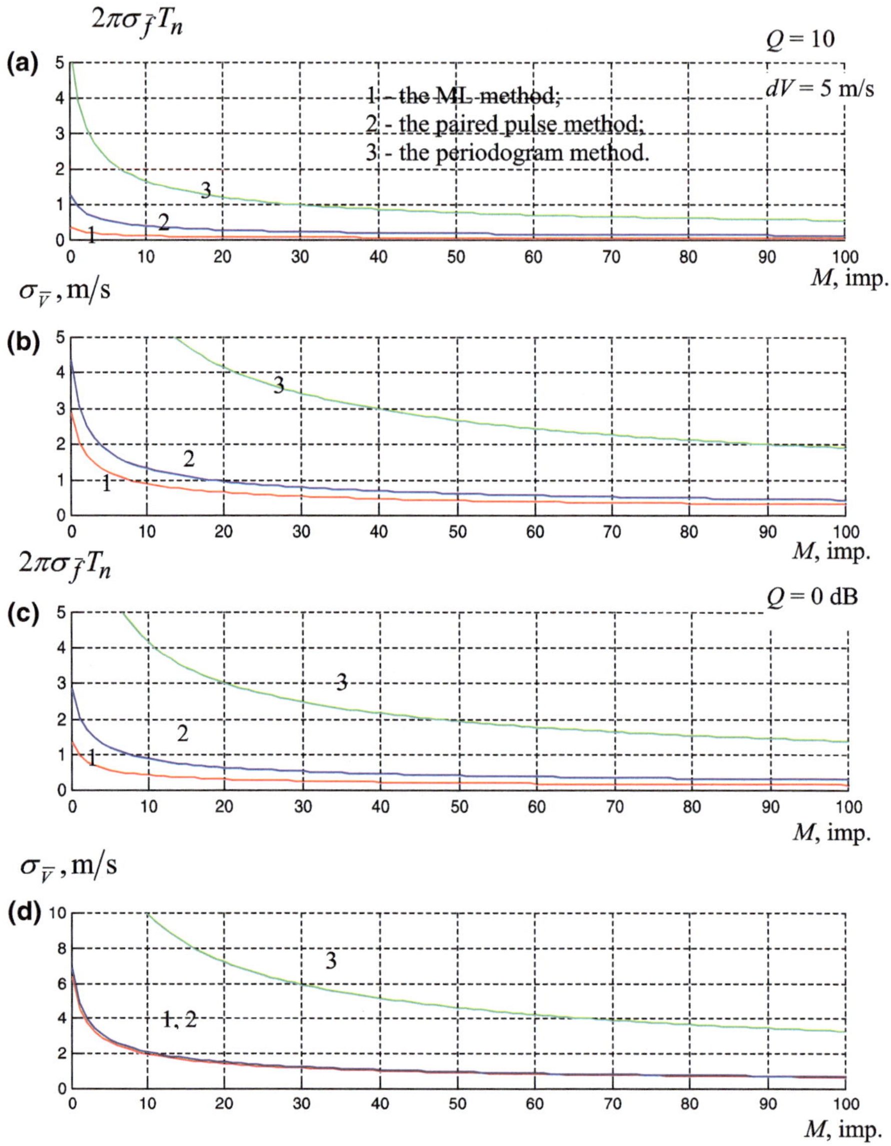

Fig. 3.3 Dependence of RMS errors of assessment of the rated average frequency of DS of MO (**a**, **c**) and the average radial velocity (**b**, **d**) from the selection length (pack volume) for the various values of SNR

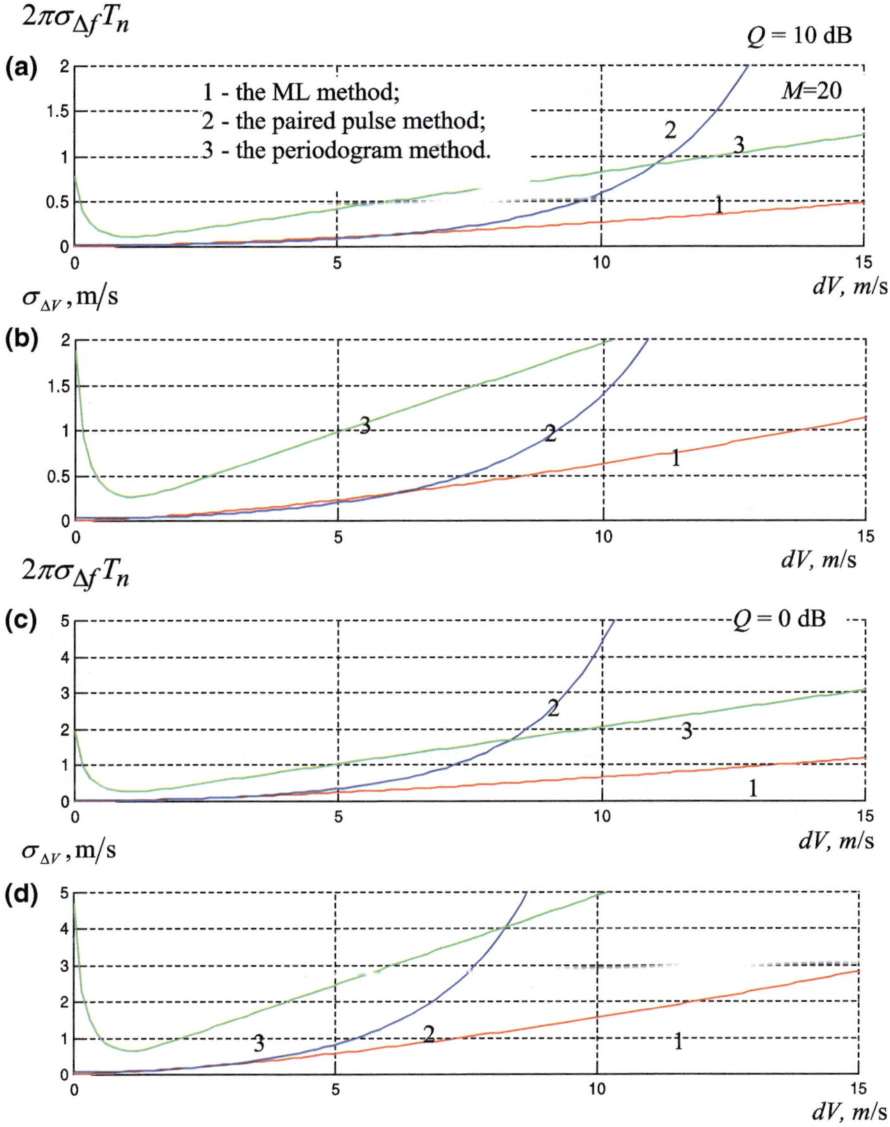

Fig. 3.4 Dependence of RMS errors of assessment of the rated width of DS of an MO signal (**a**, **c**) and pulsations of radial velocity (**b**, **d**) from intensity of atmospheric turbulence (RMS value of radial velocity of MO particles of) for the various values of SNR

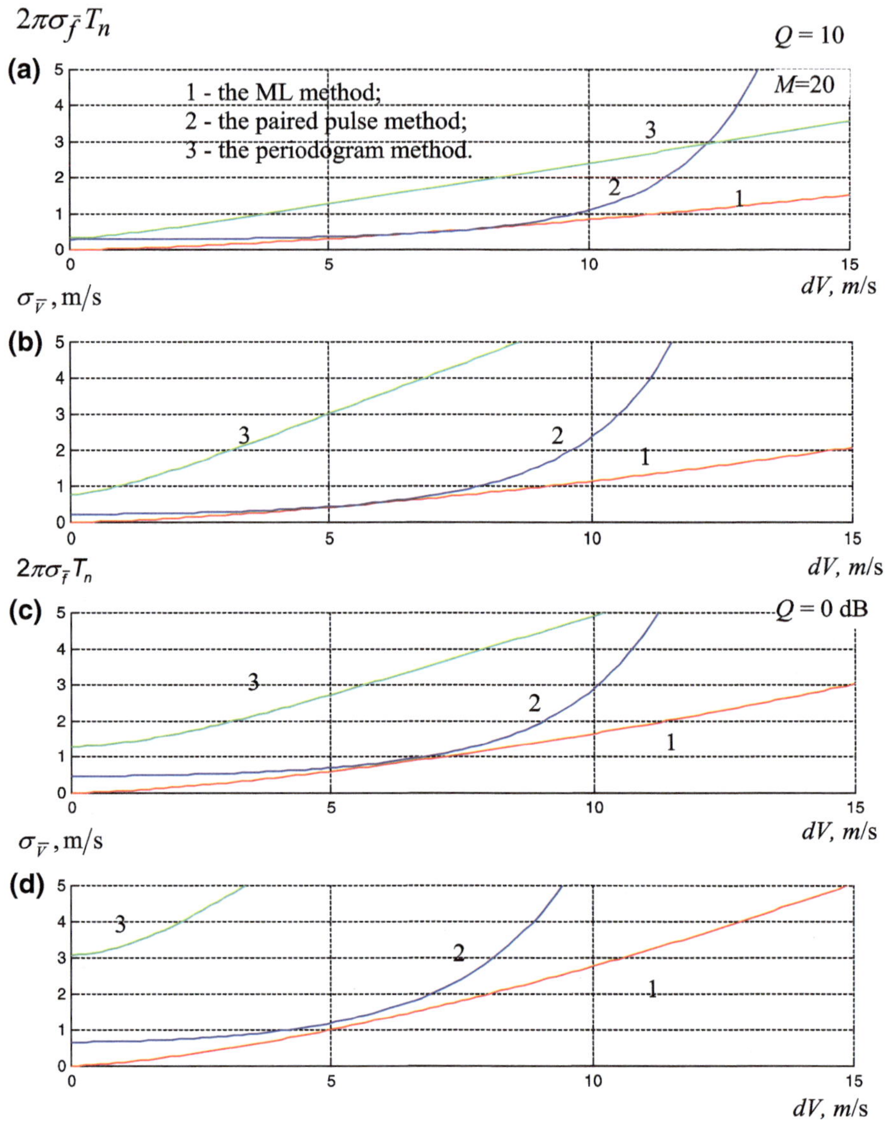

Fig. 3.5 RMS dependence on errors in assessment of rated average DS frequency of an MO signal (**a**, **c**) and average radial velocity (**b**, **d**) from intensity of atmospheric turbulence (RMS of the value of radial velocity of MO particles) for different values of SNR

At first two independent normal random sequences were generated: the forming noise $E(m)$ and the measurement noise $N(m)$. Then the noise $E(m)$ passed through the forming filter according to (2.107). The signal received at the output of the filter was combined with the noise $N(m)$, forming a vector of readings of the input mixture of the processing device.

Then the modified covariant method (Fig. 3.6) was used to assess the order and coefficients of the AR model on which the poles of PSD which has valid and imaginary part according to (2.115) and (2.116) are defined by the average frequency and width of DS of the analysed signals were calculated.

And then the array of own values and own vectors of the specified matrix is defined.

When using the MUSIC method (Fig. 3.7) on the basis of the created selection of the input mixture $X(m)$ its ACM $\underline{\mathbf{B}}_x$ and then the array of own values and own vectors of the specified matrix is defined. The number of own vectors of the ACM in a subspace of a signal is defined by minimization of the modified Akaike information criterion (A3.22) [38]

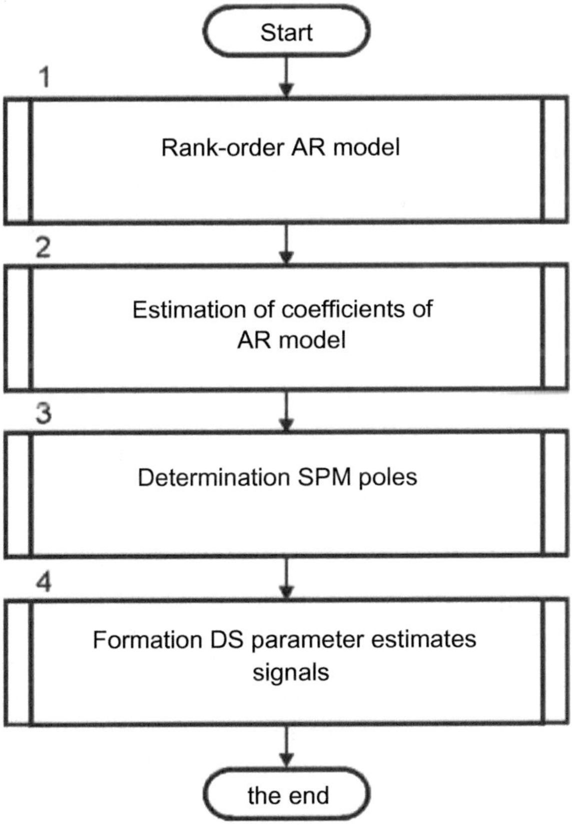

Fig. 3.6 The block diagram of the covariant algorithm of assessment of the DS parameters of signals

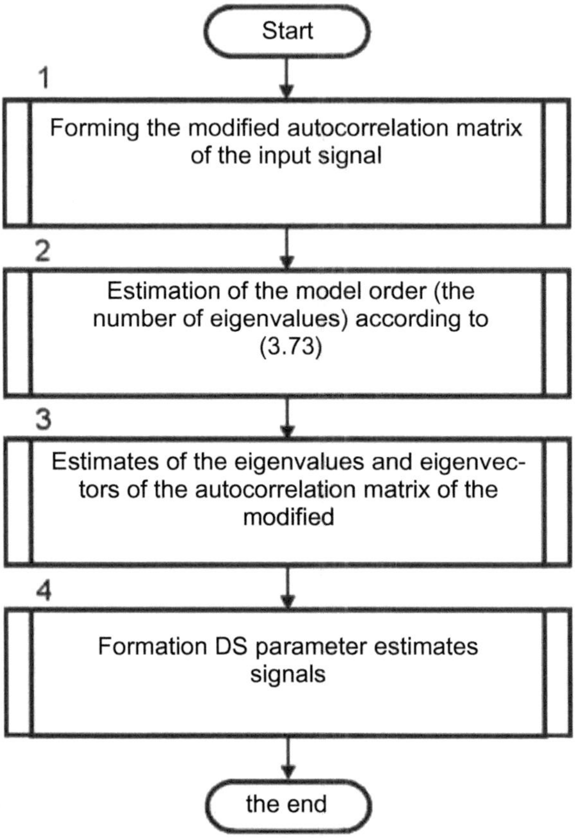

Fig. 3.7 The block diagram of the algorithm of assessment of the DS parameters of signals by the MUSIC method

$$AIC(m) = (p-m)\ln\left(\frac{\frac{1}{p-m}\sum_{i=m+1}^{p}\lambda_i}{\prod_{i=m+1}^{p}\lambda_i^{-(p-m)}}\right) + m(2p-m), \tag{3.73}$$

where $\lambda_1 > \lambda_2 > \ldots > \lambda_p$—the own values of selective ACM;
m—the number of narrow-band signals ($m < p$).

In a noise subspace ACM will have only one own vector and the unique own value ρ_k, connected with it corresponding to the dispersion σ_n^2 of management noise. The value of $\rho_k = \lambda_p$ is the ACM minimum own value. In case of mismatch of the model, that is if the rank of ACM of input mixture determined by the size M of the processed pack exceeds the number narrow-band components in a useful signal, then the matrix $\underline{\mathbf{B}}_x$ will have $(M - p)$ own vectors of a noise subspace and the corresponding own values.

Theoretically, as it follows from (3.47), all own values of noise subspace have to coincide in value, however really because of existence of errors of measurements they are just very close. The MUSIC method assumes uniform weight processing of all $(M - p)$ own vectors of a noise subspace.

After definition of noise own vector the frequencies of narrow-band components of the reflected signal are found using factorization of the polynom, which is in the denominator of the expression for DS of MO signals (3.49). Then the relative capacities P_i of narrow-band components are calculated in the course of the solution of the matrix equation

$$\begin{bmatrix} \exp(j2\pi f_1 T_n) & \exp(j2\pi f_2 T_n) & \ldots & \exp(j2\pi f_m T_n) \\ \exp(j2\pi f_1 2T_n) & \exp(j2\pi f_2 2T_n) & \ldots & \exp(j2\pi f_m 2T_n) \\ \ldots & \ldots & \ldots & \ldots \\ \exp(j2\pi f_1 mT_n) & \exp(j2\pi f_2 mT_n) & \ldots & \exp(j2\pi f_m mT_n) \end{bmatrix} \begin{bmatrix} P_1 \\ P_2 \\ \ldots \\ P_m \end{bmatrix} = \begin{bmatrix} B(0) - \lambda_{\min} \\ B(1) \\ \ldots \\ B(m-1) \end{bmatrix}.$$

The accuracy of assessment of the average frequency of the spectrum of narrow-band signals by the MUSIC method depends on the volume M of the processed pack, SNR Q and correlation properties of a signal [39]

$$\sigma_{\bar{f}}^2 = \frac{\sigma_n^2}{2M}\left[\sum_{k=1}^{p}\frac{\lambda_k}{\left(\sigma_n^2 - \lambda_k\right)^2}\left|\vec{c}^H \vec{V}_k\right|^2 \Big/ \sum_{k=p+1}^{M}\left|\vec{d}^H \vec{V}_k\right|^2\right], \tag{3.74}$$

$$\text{where } \vec{c} = \left[1 \; e^{j2\pi\bar{f}} \; \ldots \; e^{j2\pi(M-1)\bar{f}}\right]^T$$

$$\vec{d} = \frac{1}{2\pi}\frac{d\vec{c}}{d\bar{f}} = \left[0 \; je^{j2\pi\bar{f}} \; \ldots \; j(M-1)e^{j2\pi(M-1)\bar{f}}\right]^T.$$

The analysis of the expression (3.74) shows that the dispersion $\sigma_{\bar{f}}^2$ of assessment of the average frequency of DS can reach big values if some of the own values λ_k of ACM $\underline{\mathbf{B}}_x$ are close to the dispersion σ_n^2 of measurement noise. It takes place at low SNR at the input of the processing device or at observation as a part of the signal reflected by one allowed MO volume, of several narrow-band sequences with close DS [39] that is, characteristics of the assessments by the MUSIC method considerably depend on how well the breaking of a signal and noise into subspaces is carried out. The main criterion at the same time is the analysis of relative amounts of the ACM own values.

According to the flowcharts presented in Figs. 3.6 and 3.7 the SigVstat program is developed in the MATLAB system for carrying out modelling (the Annex 4). The received results are presented in Figs. 3.8, 3.9 and 3.10.

The dependence of dispersion of the assessment of the average frequency of DS of the signals reflected by MO from the selection length (pack volume) is presented in Fig. 3.8a. When modelling it was supposed that SNR Q = 3 dB, $T_p = 1$ ms, $\sigma_V = 5$ m/s (heavy turbulence), $\lambda = 3$ cm. The dependence corresponding to the ML optimum method is represented for comparison on the same schedule. The results of

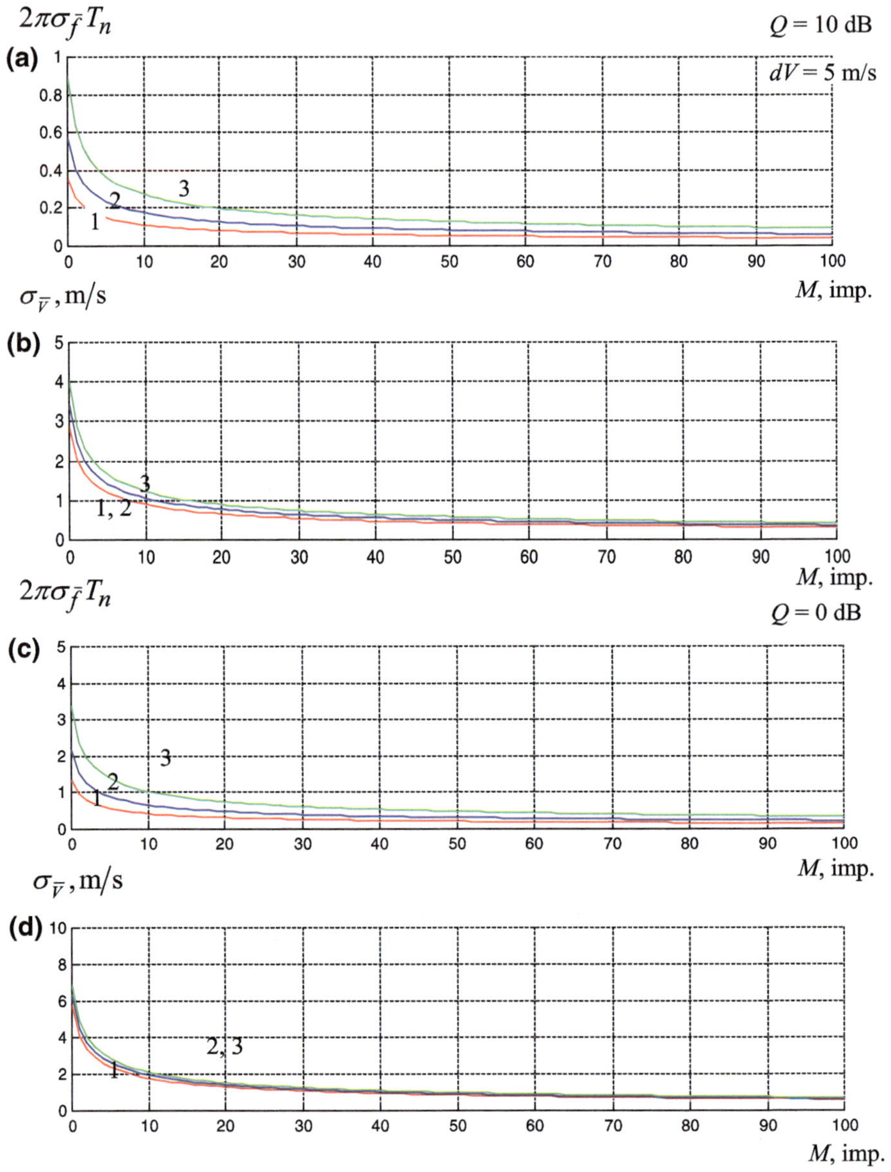

Fig. 3.8 The dependence of the RMS errors of the assessment using parametric methods of the rated average frequency of DS of an MO signal (**a**, **c**) and the average radial velocity (**b**, **d**) from the selection length (pack volume) for the SNR various values (1—the ML method; 2—the MUSIC method; 3—the covariant method)

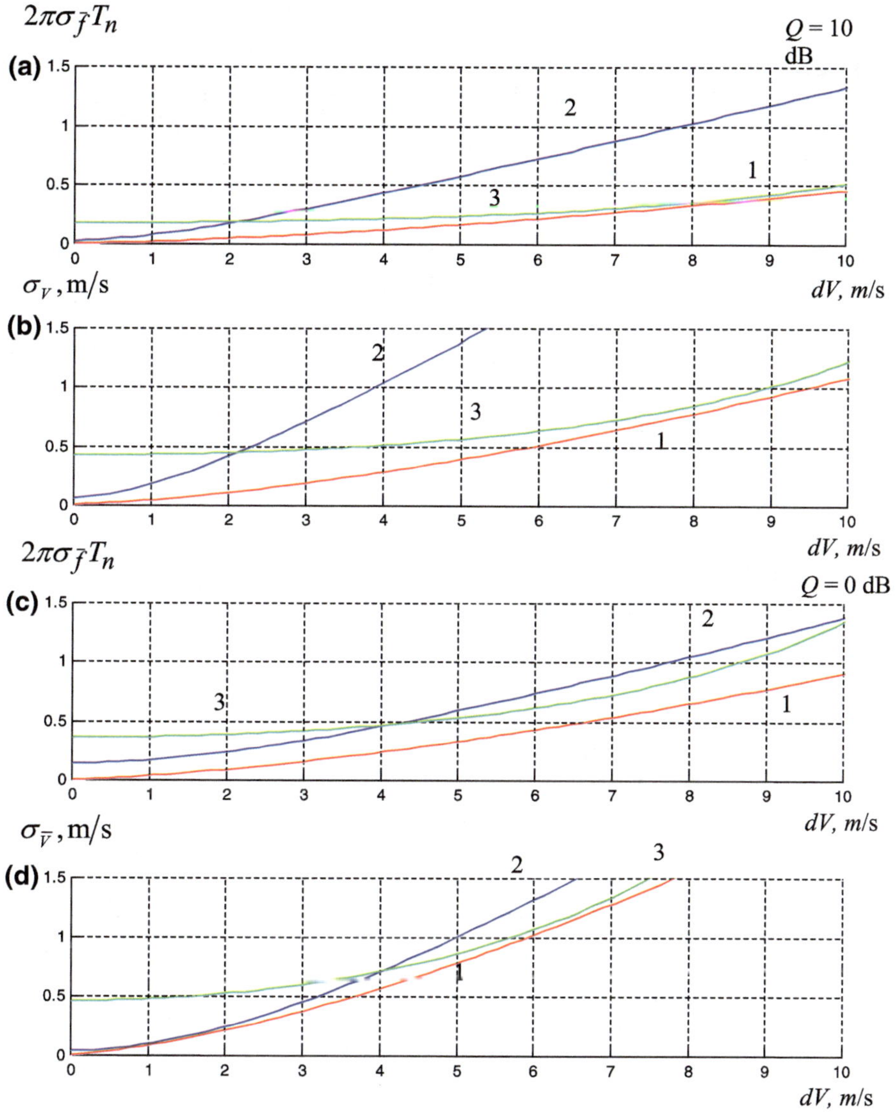

Fig. 3.9 Dependence of RMS of errors of assessment using parametric methods of rated average frequency of DS of an MO signal (**a**, **c**) and average radial velocity (**d**) from intensity of atmospheric turbulence (RMS of the value of radial RMS of MO particles) for different values of SNR (1—the ML method; 2—the MUSIC method; 3—the covariant method)

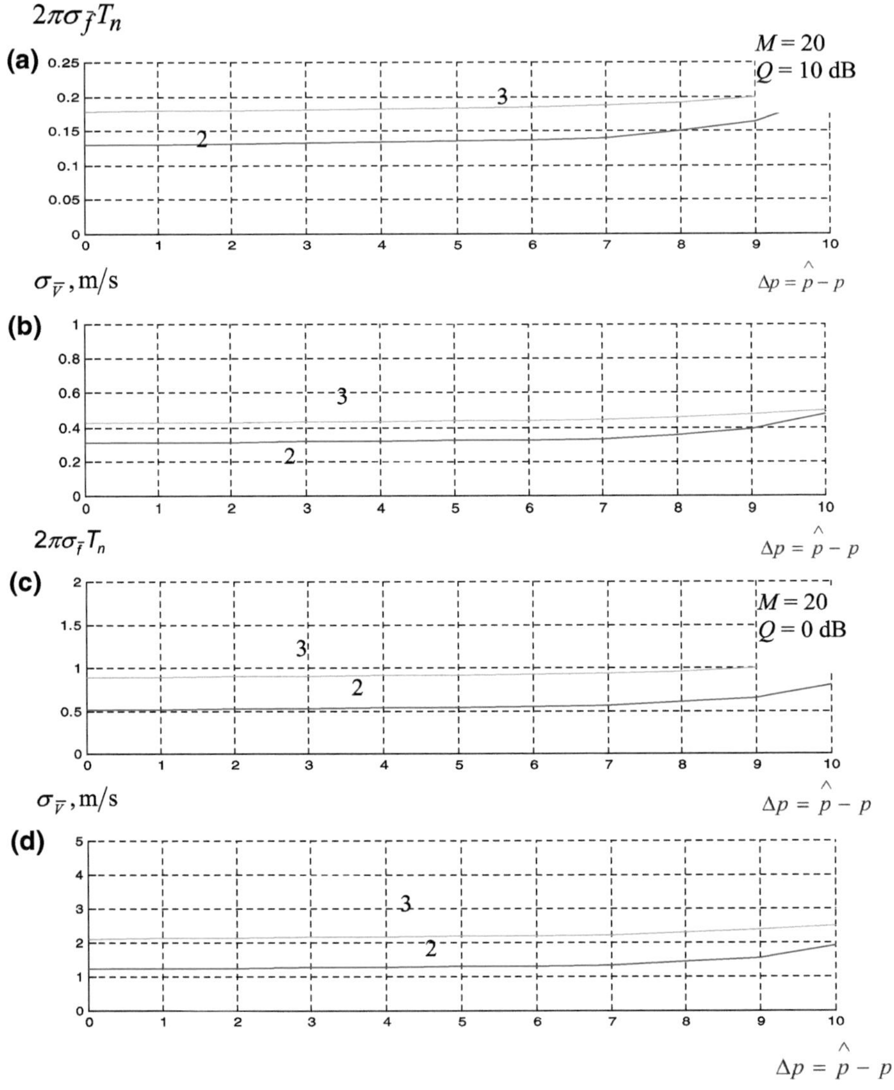

Fig. 3.10 Dependence of RMS errors of the assessment using parametric methods of the rated average frequency of DS of an MO signal of (**a**, **c**) and the average radial velocity (**b**, **d**) from the overdetermination degree of the AR model for the various values of SNR (2—the *MUSIC* method; 3—the covariant method)

modelling confirmed the inversely proportional dependence $\sigma_{\bar{f}}^2$ on M similar to the earlier received dependence for nonparametric methods (Fig. 3.3a).

The same dependence, but in other scale—for the RMSE of the assessment of pulsations of radial velocity for the allowed MO volume is presented in Fig. 3.8b. As the airborne RS should provide the accuracy of assessment of the width of a spectrum of pulsations of radial velocity of the order ±1 m/s, proceeding from results of modelling, the necessary volume of a pack will be: for the MUSIC method—10 reflected impulses, for the modified covariant method—15 reflected impulses that it is significantly less, than at using of the nonparametric methods of assessment considered above.

The dependences of the dispersion of assessment of the width of a Doppler range of the signals reflected by MO and the RMSE of the assessment of pulsations of radial velocity from a pack volume received at smaller SNR are given in Fig. 3.8c and d (Q = −3 dB). In this case the minimum admissible volume of a pack is about 40 impulses (for both considered methods).

The dependences of the dispersion of assessment of the average frequency of DS of the reflected signals and also average value of radial velocity of particles of the allowed MO volume on intensity of the atmospheric turbulence determined by the RMS value of the radial velocity are presented in Fig. 3.9. These dependences confirm the earlier made conclusion that the accuracy of the assessment of the corresponding parameters by the MUSIC method focused on processing of rather narrow-band signals with widening of a signal spectrum worsens much quicker, than for the modified covariant method.

In the solution of the problem of detection of the zones of intensive atmospheric turbulence forming the reflected signals with wide DS, preference should be given to the modified covariant method. This conclusion is especially relevant at installation of airborne RS on the high-speed carrier.

Here it should be noted that the assessments of the moments of DS of MO signals by parametric methods significantly depend on adequacy of the assessment of the AR model order. If the model order is overestimated that corresponds to an inconsistent model, then the accuracy of assessment worsens, especially by the MUSIC [31] method. It is confirmed by the results of modelling presented in Fig. 3.10.

As the MUSIC method is based on the analysis of own values and own vectors of the ACM of the accepted mixture of a useful signal and noise, the analysis of dependence of accuracy of the assessment of the corresponding own values and own vectors of the ACM from the volume of the accepted pack of signals (selection length) is of interest. At the assessment of the ACM own values the necessary volume of a pack is out of the condition [51]

$$M \leq 2\lambda_i^2 \overline{(\hat{\lambda}_i - \lambda_i)}^{-2},$$

where $\hat{\lambda}_i$—the assessment of the i-th own value λ_i of ACM.

At replacement of the maximum own value of ACM with its top side we will receive that with the error $\Delta\lambda_i$ the necessary volume of a pack is $M < 2p/\Delta\lambda_i$. Here $\Delta\lambda_i$—the accuracy of the assessment of i-th own value λ_i. For example, at dimension of ACM $p = 10$ and the accuracy of assessment $\Delta\lambda_i = 10\%$ the upper bound of a necessary pack does not exceed 200, and at $\Delta\lambda_i = 100\%$ (that is often admissible in practice as corresponds to the error of assessment of power of spectral components the equal 3 dB) the selection up to 20 readings long is necessary. The required pack volume for assessment of own vectors $\vec{V}_i$ is a bit bigger.

3.1.3.2 The Analysis of Influence of MO Characteristics and RS Parameters on the Accuracy of Assessment of the Moments of a Doppler Spectrum

Besides the errors determined by the used processing method the accuracy of assessment of the moments of a Doppler spectrum of the reflected signals is influenced significantly by MO characteristics and airborne RS parameters.

In particular, besides turbulence, the relative movement of reflectors in the radial direction leading to the DS widening can be caused by some other meteorological phenomena (in particular, presence of longitudinal and cross WS) and also gravitational falling of reflectors of the different size. The presence of several factors independently changing the relative velocity of the reflector leads to the fact that the velocity of the separate reflector can be written down in the form of the superposition of several components

$$V_n = \sum_i V_{ni}, \tag{3.75}$$

where V_{ni}—the n-the reflector velocity component caused by influence of the i-th factor.

In case of independence and random nature of these factors the dispersion of a spectrum of velocities of reflectors also represents the superposition of components

$$\Delta V^2 = \sum_i \Delta V_i^2, \tag{3.76}$$

where ΔV_i^2—the component of the dispersion of a spectrum of velocities of reflectors as a part of the allowed MO volume, caused by influence of the i-th factor.

The presence of the additional factors changing radial velocities of reflectors and widening a spectrum of velocities (therefore, also DS of the reflected MO signals), worsens the turbulence assessment accuracy on the RMS spectrum width.

The dispersion of a spectrum of velocities can be presented in the form the sum [9]

$$\Delta V^2 = \Delta V_t^2 + \Delta V_{ws}^2 + \Delta V_g^2, \tag{3.77}$$

where ΔV_t^2—the contribution of small-scale turbulence caused by dispersion of velocities of reflectors in relation to the average velocity $\overline{V}$;
ΔV_{ws}^2—the WS contribution caused by change of average speed $\overline{V}$ of within the allowed volume;
ΔV_g^2—the dispersion of the spectrum of velocities of gravitational falling (taking into account the projection to the direction of sounding).

The presence of WS leads to the fact that various HM which are a part of the allowed volume receive various increments of velocity, that is there is widening of the spectrum of velocities.

For narrow enough AS DP ($\Delta\alpha$, $\Delta\beta$ about several degrees) on limited intervals of heights the change of the projection of wind speed with height can be approximated by the linear function [52]

$$V_w(z) = V_{w0} + k_w z, \tag{3.78}$$

where k_w—a vertical gradient of wind speed;
V_{w0}—wind speed at the flight altitude of the RS carrier in the location of MO;
z—the height counted from the plane of the movement of the RS carrier.

As it is specified in [53–56], generally the values of a gradient of wind speed change with height. The whole interval of heights from the Earth's surface to 12 km, within which MO can be observed, can be broken into several layers with the differing values of k_w. In the layer of 0.3–1 km k_w is 7–9 m/(s·km), seasonally. At the heights of 1.2 km the gradient is concluded in the range of 5–6.5 m/(s·km). At the heights of 2–6 km k_w poorly changes with height and has small dispersion about 4 m/(s·km). At the heights of 7–12 km of value of the gradient increase again up to 5–7 m/(s·km). Usually the value $k_w = 5$ m/(s·km) is accepted in calculations [56].

Let the direction of wind speed make the angle α_w with the direction of air velocity of the RS carrier. As the height $z \approx r \sin\beta$, where r—the distance to the centre of the corresponding allowed volume, the expression (3.78) can be rewritten in the form

$$V_w = V_{w0} + k_w r \sin\beta. \tag{3.79}$$

The radial component of velocities of reflectors caused by influence of WS is equal to

$$V_{wr} = V_w \cos(\alpha - \alpha_w)\cos\beta = (V_{w0} + k_w r \sin\beta)\cos(\alpha - \alpha_w)\cos\beta. \tag{3.80}$$

Let us express the angles α and β through φ and ψ taking into account (2.1) and expand V_{wr} in Taylor series, being limited to linear members

$$V_{wr} = V_0' - \varphi V_{\langle}' - \psi V_{‰}',$$

where $V_0' = (V_{w0} + k_w r \sin\beta_0)\cos(\alpha_0 - \alpha_w)\cos\beta_0$,

$$V'_{\langle} = (V_{w0} + k_w r \sin\beta_0)\sin(\alpha_0 - \alpha_w),$$
$$V'_{‰} = (V_{w0}\sin\beta_0 - k_w r\cos 2\beta_0)\cos(\alpha_0 - \alpha_w).$$

The following Doppler frequency corresponds to the velocity V'_0

$$\bar{f}' = 2V'_0/\lambda = 2\lambda^{-1}(V_{w0} + k_w r\sin\beta_0)\cos(\alpha_0 - \alpha_w)\cos\beta_0, \tag{3.81}$$

which is the average frequency of DS of the reflected signal.

As the wind speed can have any sign, presence of wind can lead both to increase, and to decrease of average frequency of a spectrum. Besides, the more the AS DP axis deviates from the direction of the movement of the RS carrier of radar station, the less is the value of Doppler shift (3.81).

Now we will estimate the contribution of WS to the width of the spectrum of the reflected signals. If the density of reflectors within the allowed volume is constant, then the distribution of power of the reflected MO signals, depending on the angles α and β, will be described by AS DP square.

If we approximate a square of DP by the Gaussian function

$$G^2(\alpha, \beta) = \exp\left(-\frac{(\alpha - \alpha_0)^2}{2\Delta\alpha^2}\right)\exp\left(-\frac{(\beta - \beta_0)^2}{2\Delta\beta^2}\right)$$

then

$$p(\beta) = \int_0^{2\pi} G^2(\alpha, \beta)d\alpha = \exp\left(-\frac{(\beta - \beta_0)^2}{2\Delta\beta^2}\right) \approx \exp\left(-\frac{\beta^2}{2\Delta\beta^2}\right), \tag{3.82}$$

For the distribution of the form (3.82) the rated spectrum of velocities will also be defined by the Gaussian function repeating the AS DP form in a certain scale

$$S(V_w) = \exp\left(-\frac{\left(V_w - \overline{V_w}\right)^2}{2\Delta V_{ws}^2}\right), \tag{3.83}$$

where $\overline{V_w} = V_{w0}$;

$$\sigma_{V-‰} = \frac{\Delta V_{wr}}{2\sqrt{2\ln 2}} = \frac{k_w r\Delta\beta}{2\sqrt{2\ln 2}}. \tag{3.84}$$

$\Delta V_{wr} = k_w r\Delta\beta$—the width of distribution of radial wind speed on the level of half power

The feature (3.84) is the dependence of the RMS spectrum width on distance r from which the reflected signal is accepted. It is explained by the fact that depending

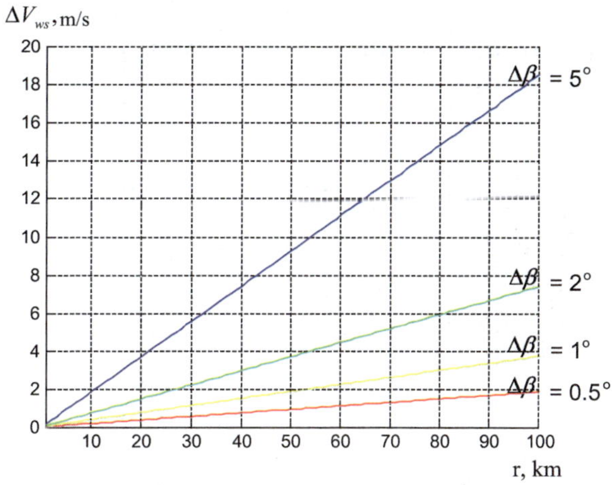

Fig. 3.11 The dependence of the width of spectrum of HM velocities caused by s vertical WS depending on range for various values of the width $\Delta\beta$ of DP on an elevation angle ($\Delta\beta = 0.5$, 1, 2, 5°)

on distance of DP the sites of MO with various differences of velocities on height are irradiated.

The dependence of the width of the spectrum of HM velocities caused by a vertical WS depending on a range for various values of the width $\Delta\beta$ of DP on an elevation angle calculated by the formula (3.84) by means of the Sigma_v.m program is given in the Fig. 3.11.

Generally, when there is some nonzero angle between the direction of wind and the direction of sounding, the connection between a vertical gradient of wind speed and the RMS width of the spectrum of HM velocities (3.84) has the form

$$\Delta V_{ws} = \frac{k_w r \Delta\beta}{2\sqrt{2\ln 2}} \cos(\alpha - \alpha_w) \cos\beta. \tag{3.85}$$

The contribution ΔV_g of gravitational velocity of falling of reflectors is caused by the distinction of vertical velocities of HM falling under the influence of gravity and air resistance. Gravitational velocities of falling of HM are unambiguously defined by their size, phase state and the ratio of density of HM and air [57]. For water drops the established speed of their falling in motionless air is defined by the formula [58]

$$V_g(R) = V_\infty\left[1 - \exp\left(-a_g R\right)\right], \tag{3.86}$$

rather well describing the available experimental data about velocities near the Earth's surface, except for very large drops.

In [59] using the results of the analysis of experimental materials it is claimed that gravitational velocities of falling of liquid drops in motionless air do not exceed 10 m/s. The smaller velocities of falling not exceeding 2–2.5 m/s are characteristic of snowflakes and crystal particles.

Besides, as it is noted in [60], there is an empirical dependence between the average gravitational velocity of falling of HM and R reflectivity—the so-called "V-Z ratio"

$$\overline{V}_g = A_g Z^{b_g}, \tag{3.87}$$

where A_g and b_g—the coefficients depending on MO microstructure.

In [61] it is noted that the values A_g lie in the range of 2.6–3.84, and b_g—in the range 0.071–0.114. The dispersion of the values calculated on various V-Z ratios does not exceed ±1 m/s.

The contribution of gravitational velocity of falling of reflectors will be shown at big elevation angles

$$\Delta V_g = \Delta V_g^0 \sin\beta,$$

where ΔV_g^0—where—the RMS width of the spectrum of gravitational velocities at vertical sounding, for drops of rain $\Delta V_g^0 \approx 1$ m/s [62].

As the gravitational velocity of falling of the reflector depends on its size [63]

$$V_{ng}(R) = a_R\big[1 - \exp(-b_R R)\big] - \gamma_R R, \tag{3.88}$$

where $a_R = 18.6711$; $b_R = 635.825$; $\gamma_R = 2289.758$, $[R] = \text{m}$, then ΔV_g^0 is determined only by distribution of MO particles on sizes.

Here it should be noted that the width of a DS of the reflected signals is the function of not only the MO parameters characterizing the distribution of elementary reflectors in space and on velocities but also and of the RS parameters (width and velocities of scanning by the AS DP, stability of the carrier frequency of RS PS, bandwidth of the RS receiver, impulse duration, etc.).

Let us consider the impact of scanning of the antenna on the widening of a DS of the accepted signals [52, 62, 64, 65]. Let the reflectors included in MO be "frozen" (motionless). At motionless AS DP the signal reflected by them will have the constant amplitude and a phase for each certain allowed volume. At scanning DP two timepoints, t and t + τ, correspond to two allowed volumes equidistanted from radar station but displaced on the angle $\Omega_a \tau$, where Ω_a—the angular velocity of scanning. The part of elementary reflectors which were irradiated in the timepoint t also remains within a beam of a radar station and at the timepoint t + τ. However, some reflectors are not irradiated any more, but again appeared ones are irradiated. Thus, there is an updating of structure of the irradiated elementary reflectors as a part of the allowed volume.

The total number of the irradiated reflectors at the moments t and t + τ is almost identical, but their structure is different: the more τ, the less continuity is in this structure. At AS DP shift on the angle equal to its width, the complete updating of structure of reflectors occurs. In this case the component of width of a spectrum of fluctuations of the reflected signals is equal to

$$\Delta f_{sc} = \frac{1}{2\tau_{sc}} = \frac{1}{2t_{\Delta\alpha}} = \frac{\Omega_a}{2\Delta\alpha} = \frac{F_\alpha}{2\Delta\alpha T_v}, \tag{3.89}$$

where τ_{sc}—the interval of correlation of the fluctuations caused by DP scanning;
$t_{\Delta\alpha}$—the time of the analysis of a resolution element on an azimuth;
Ω_a—the angular speed of scanning;
$\Delta\alpha$—the AS DP width on an azimuth;
Φ_α—the angular size of the sector of scanning;
T_v—the view period.

At $\Phi_\alpha = 180°$, $T_v = 50$ s, $\Delta\alpha = 2.4°$ (the Sect. 1.4.2) the width of the spectrum of the reflected signals, caused by scanning by AS DP, will be according to (3.117) $\Delta f_{sc} = 0.75$ Hz.

The instability of frequency of RS PS also leads to fluctuations. As shown in [66] in order to neglect them, it is necessary that $df_0 << (2\tau_u)^{-1}$, where τ_u—the duration of RS PS. Thus, the fluctuations of the reflected signal caused by instability of the carrier frequency of the RS transmitter do not depend on its operating frequency, and are completely defined by the PS duration.

However, at the modern level of development of R equipment such high requirements are imposed to the PS carrier frequency stability that fluctuations of the signals reflected by the allowed MO volume because of instability of frequency are significantly lower than the fluctuations caused by other reasons and can be neglected.

The accuracy of assessment of the DS parameters of the reflected signals is also influenced by the RS AS radiotransparent dielectric radome which brings additional phase shift in the accepted EMW [67]. The fluctuations of the brought phase shifts and the transfer coefficient on an aperture of the RS AS are caused by a radome form lead to distortion of AS DP (change of width of the main lobe, maximum shift, increase in LSL). The extent of influence on AS DP on a radome form (aerodynamic—cone-shaped for supersonic velocities or spheroidal for subsonic velocities) and radome wall designs (single-layer or multilayered) therefore the two-dimensional dependence of the brought phase shift of EMW on the azimuth and elevation angle characterizing the direction on EMW source in relation to the plane of an aperture of RS AS cannot be presented in the analytical form. The specified phase shifts can be measured in the course of natural experiments and are written down into the memory device (MD). Then these data are used in OBC for compensation of amplitude and phase distortions.

The technical parameters of radar station influence as well the accuracy of assessment of the average frequency of an MO DS. In particular, at approximation of the signal reflected by the allowed volume by a simple Markov process with the correlation function of the form

$$B_s(\tau) = \sigma_s^2 \rho(\tau) = \sigma_s^2 e^{-\gamma|\tau|}$$

and at SNR $Q > 1$ the potential accuracy of the assessment of the average frequency of a DS of an MO signal is defined by the ratio [68]

$$\sigma_{\bar{f}}^2 = \frac{M^2}{16\pi^2 Q^2 T_n^2 \sum_{i=1}^{M-1} (M-i) i^2 \rho^2(T_n)}. \tag{3.90}$$

From this ratio it is seen that the accuracy of measurement of the average Doppler frequency of an MO signal does not directly depend on the RS PS duration (the Sect. 2.4.3.2). However, with the growth of τ_u, the amount of the allowed volume and, respectively, HM EER filling this volume increase. For this reason even at the fixed energy of RS PS the SNR value Q will increase and, respectively, the value of $\sigma_{\bar{f}}^2$ will decrease. On the other hand, with temporal not stationarity and spatial heterogeneity of MO with the growth of τ_u not only the resolution of radar station on range δr, but also frequency assessment accuracy $\bar{f}$ will worsen. It is obvious that the choice of PS duration should be made proceeding from the condition of ensuring the required resolution of a radar station on range.

The expression (3.90) confirms as well the statement given above about the significant influence on the accuracy of assessment of the average frequency $\bar{f}$ of the value of the coefficient of interperiod correlation $\rho(\tau)$, determined by the width Δf of the spectrum of low-frequency fluctuations of an MO signal. This influence is especially strongly shown at big durations of the reflected signals (at $MT_n > 40$–50 ms).

3.1.3.3 The Analysis of Stability of the Assessment of the Moments of a Doppler Spectrum

According to the theory of linear systems [3, 13], the property of stability of a system is that limited input signals create limited signals at the output. Stability of parametric spectral assessments consists of two components: stability of the corresponding model forming filter and statistical stability.

Stability of the filter is provided with the choice of the AR model parameters. The necessary and sufficient condition of stability of the filter consists in the minimum phase of the polynom $A(z)$ in the denominator of an z-image of the fractional and rational transfer function (TF) (2.113) of the linear forming AR model filter that is equivalent to the condition of all roots of the polynom $A(z)$ (TF poles) being in a single circle on z-plane: $|z_k| < 1,\ k = 1, \ldots, p$. Otherwise there is not stationarity of an output signal in the form of the unlimited growth of its dispersion that contradicts real physical data.

Statistical stability characterizes the possibility of use of these or those methods of spectral estimation during the work with set final records of data of any form and also their sensitivity to rounding errors in the course of calculations. The insufficient data capacity determined by the dynamic range of input signals, and increase in noise of quantization at performance of arithmetic operations (summation, multiplication) can result in statistical instability of the used computing algorithm.

Let us consider the assessment of the vector of the AR coefficients $\vec{a}$ on input selection $\vec{x}$ with the volume of $M \times 1$ (taking into account (3.35))

$$\underline{\mathbf{B}}\vec{a} = \vec{P},$$

where $\underline{\mathbf{B}} = [B(i, j)]$—the ACM of an input signal,

$$B(i, j) = \sum_{m=0}^{M-1-p} \left[x^*(m-i)x(m-j) + x^*(m-p+i)x^*(m-p+j)\right];$$

$$\vec{P} = \left[\rho_p\ 0 \ldots 0\right]^T.$$

In case the ACM of an input signal is estimated with the errors caused by insufficient capacity of input data and rounding of results of multiplication and accumulation, its assessment will differ from the true value by quantity

$$\underline{\Delta\mathbf{B}} = \hat{\underline{\mathbf{B}}} - \underline{\mathbf{B}}.$$

The errors of assessment of ACM in the result of calculations will lead to an essential error at estimation of the AR coefficients and, as a result, to errors in the assessment of the average frequency and width of a DS of the signals reflected by MO.

Let us consider that elements of the matrix $\underline{\Delta\mathbf{B}}$ are independent, equally distributed random values, such that

$$E\{\Delta B(i, j)\} = 0, \quad E\left\{\Delta B(i, j)\Delta B^*(k, l)\right\} = \sigma^2\delta_{ik}\delta_{jl}, \tag{3.91}$$

where σ^2—dispersion $\Delta B(i, j)$;
δ_{ik}—Kronecker delta.

This model allows to define the upper bound of the ineradicable errors caused by the finite capacity of representation of input processes and also errors of calculations.

The correlation matrix of the vector $\vec{P}$, corresponding to the rated ACF of the forming noise has the form

$$\underline{\mathbf{B}}_P = E\left\{\vec{P}\vec{P}^*\right\} = E\left\{\hat{\underline{\mathbf{B}}}\vec{a}\vec{a}^*\hat{\underline{\mathbf{B}}}^*\right\}$$
$$= \underline{\mathbf{B}}E\left\{\vec{a}\vec{a}^*\right\}\underline{\mathbf{B}}^* + E\left\{\underline{\Delta\mathbf{B}}\vec{a}\vec{a}^*\underline{\mathbf{B}}^*\right\} + E\left\{\underline{\mathbf{B}}\vec{a}\vec{a}^*\underline{\Delta B}^*\right\} + E\left\{\underline{\Delta\mathbf{B}}\vec{a}\vec{a}^*\underline{\Delta\mathbf{B}}^*\right\}.$$

As the elements of the matrix $\underline{\Delta\mathbf{B}}$ and the vector $\vec{a}$ are independent random variables, then

$$\underline{\mathbf{B}}_P = \underline{\mathbf{B}}E\{\vec{a}\vec{a}^*\}\underline{\mathbf{B}}^* + E\{\underline{\Delta\mathbf{B}}\vec{a}\vec{a}^*\underline{\Delta\mathbf{B}}^*\}. \tag{3.92}$$

According to criterion of the minimum of RMSD

$$\mathrm{tr}\left[\underline{\mathbf{B}}_P\right] \to \min,$$

where $\mathrm{tr}[\underline{\mathbf{B}}_P]$—the operation of calculation of a trace of the matrix.

Using the condition of orthogonality (3.91), it is possible to show that the matrix is $E\{\underline{\Delta \mathbf{B} \Delta \mathbf{B}^*}\} = M\sigma^2\underline{\mathbf{I}}$-diagonal. Substituting the specified value in (3.92), we receive

$$\mathrm{tr}[\underline{\mathbf{B}}_P] = M(1 + \sigma^2\mathrm{tr}[\underline{\mathbf{B}}_a]).$$

Considering the fact that because of the finite capacity of data presentation and coefficients there is an increase in power of noise by 0.1 dB (permissible value for a path of preprocessing of airborne RS radar stations [69]), it is possible to estimate σ^2

$$M\sigma^2\,\mathrm{tr}[\mathbf{B}_a] = k_3\,\mathrm{tr}[\mathbf{I}] = k_3 M\,,$$

where $k_3 = 0.023$—the coefficient corresponding to increase in power by 0.1 dB.

From here we find that

$$\sigma^2 = \frac{k_3}{\mathrm{tr}[\underline{\mathbf{B}}_a]} = \frac{k_3}{\mathrm{tr}[\underline{\mathbf{B}}_s + \underline{\mathbf{B}}_n]} = \frac{k_3}{\mathrm{tr}[\underline{\mathbf{B}}_n]\left(\dfrac{\mathrm{tr}[\underline{\mathbf{B}}_s]}{\mathrm{tr}[\underline{\mathbf{B}}_n]} + 1\right)}.$$

In the assumption that the dynamic range of signals is 2^k, the maximum own value of the matrix $\underline{\mathbf{B}}^{-1}$, equal to $\lambda_{\mathrm{min}}^{-1}$, is defined by the value 2^{2k+1}. Designating $\mathrm{tr}[\underline{\mathbf{B}}_s]/\mathrm{tr}[\underline{\mathbf{B}}_n] = Q$—SNR and an input of the system of processing, and considering that $\mathrm{tr}[\underline{\mathbf{B}}_n] = M\lambda_{\mathrm{min}}$, we receive

$$\sigma^2 = \frac{k_3}{M\lambda_{\mathrm{min}}(Q+1)} = \frac{2^{2k+1}k_3}{M(Q+1)}. \tag{3.93}$$

Defining the number of bits k of presentation of the processed data, from (3.93) we have

$$k = 0{,}5\left[2\log_2\sigma + \log_2 M + \log_2(Q+1) - \log_2 k_3 - 1\right]. \tag{3.94}$$

As it is clear from (3.94), the number of bits of data presentation and coefficients is proportional to the logarithm to the base 2 of a number of readings in the processed pack and in the same way depends on SNR at input and admissible dispersion of an error of calculations. In Fig. 3.12 the dependence k on pack volume at the fixed dispersion of an error of calculations $\sigma^2 = 1$ and SNR various values at input of the system of processing is presented.

Besides, the extent of influence of capacity of data presentation and coefficients significantly depends on the used algorithm of spectral estimation. As a result of comparison of indicators of the most widespread parametric methods of spectral

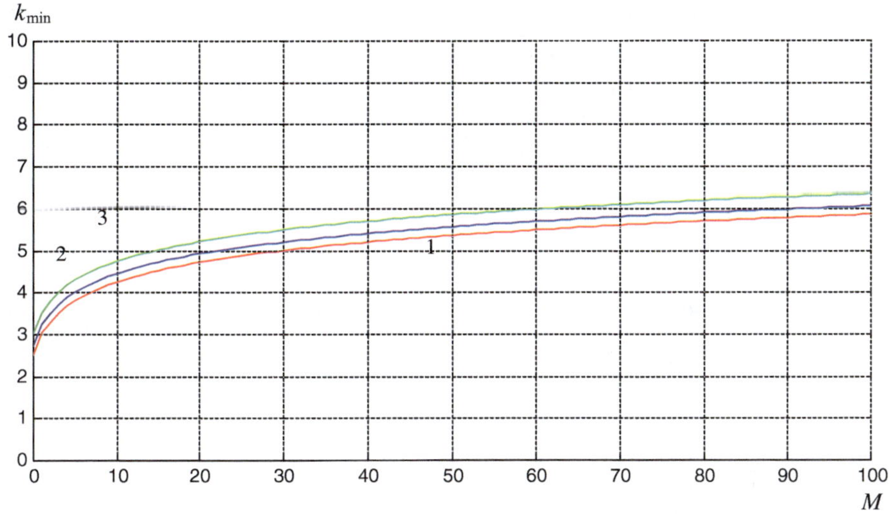

Fig. 3.12 The dependence of the minimum number of categories of coefficients of the processing device on a pack volume at the different values of SNR (1—$Q = -3$ dB; 2—$Q = 0$ dB; 3—$Q = 3$ dB)

estimation (the Burg method, the covariant method and the MUSIC method) received as a result of test calculations [70], it turned out that the Burg method is steadier in the specified sense. It is one of the few AR algorithms which successfully work with data of usual accuracy (32-bit). At the same time the realization of the covariant algorithms on the basis of arithmetics of usual accuracy leads to relative errors of assessment of the DS parameters of MO signals equal to units and even tens of percent. The reason is that these algorithms are based on matrix algebra which is very sensitive to the capacity of the processed data [71]. When using the arithmetics of the doubled accuracy (64 bits) the relative error of covariant algorithms obviously is less than one percent, which is sufficient for practical purposes.

3.1.3.4 The Analysis of Computational Efforts at Realization of Parametric Methods of the Assessment of the Moments of a Doppler Spectrum

At assessment of the moments of DS the use of two types of the digital signal processing (DSP) is possible: processing in hard real time and processing with accumulation of readings of input signals in the RSP RAM.

When processing in hard real time the DSP device should process an input signal for the sampling interval determining the rate of receipt of information. For this purpose it is often necessary to parallelize processing that can demand the use of a large number of alarm processors. Processing in real time is characteristic of consecutive (recursive) methods of assessment.

When processing with accumulation the readings of an input signal during certain time period are recorded in the RSP RAM, and then are processed. During processing there is accumulation of the new block of readings of an input signal. The delay time of a signal is defined by time of accumulation and time of processing. Processing with accumulation of signals is characteristic of block methods of processing.

In the analysis of the signals reflected by the allowed MO volume for the purpose of assessment of the moments of DS the preference should be given to processing methods with accumulation. This decision is caused by the following reasons:

- the volume of packs of the reflected signals is small because of high speeds of flight of modern AV and small ranges of detection of dangerous MO;
- interperiod (Doppler) processing demands a delay of the reflected signals for a certain number of the periods of repetition;
- when using block methods, the assessments of parameters are formed with the velocities determined by the necessary size of the processed block.

As the solution of the problem of detection of dangerous MO zones is implemented in the form of one or several operation modes of an airborne RS ("Meteo", "Turbulence", "Wind shear"), for the choice of the most preferable method of assessment of the moments of a DS of the reflected signal, it is necessary to compare the considered processing methods in terms of requirements to the main characteristics of RSP which is a part of an airborne RS (memory volume, volume of calculations, etc.).

The ML method is based on the solution of the problem of search of a global extremum of nonlinear functionality in multidimensional space. The essential complexity of implementation of this procedure caused a limited application of the ML method in practice.

The widely used now methods of spectral estimation are based on the analysis of own values of matrixes (MUSIC, ESPRIT, the minimum norm, etc.), assume complex transformations of the previously estimated ACM of the accepted signals, the assessment of own values and vectors of this matrix and also search of roots of polynoms of high orders that imposes strict requirements to temporal and computing resources of an airborne RS [29, 31, 39].

Proceeding from the above-stated information, preference should be given to realization of the modified covariant method. Its use allows to reduce significantly computational efforts in comparison with the Burg algorithm, especially in the conditions of small selections of the accepted signals (the Table 3.1). In particular, at M = 6 and p = 4 the number of operations of addition and multiplication of complex numbers necessary for the modified covariant algorithm is equal to 37 and 42 respectively against 67 and 70 for the Burg algorithm.

As at processing it is necessary to consider the phase ratios of signals, all processing should be carried out in the complex form $s(i) = s_c(i) + js_s(i)$, where $s_c(i)$ and $s_s(i)$—inphase (cosine) and quadrature (sinus) signal components.

Table 3.1 Computational efforts of the assessment of the moments of a DS of MO signals

Complex operations	The Burg method	Covariant algorithm	MUSIC method
Addition	$4Mp - \frac{3}{2}p^2 + \frac{M}{2} - 2p$	$\frac{1}{6}p^3 + Mp + \frac{1}{2}p^2 + M - \frac{8}{3}p - 1$	$\sim M^3$ [3]
Multiplication	$4Mp - \frac{3}{2}p^2 + \frac{M}{2} - \frac{5}{4}p$	$\frac{1}{6}p^3 + Mp + \frac{1}{2}p^2 + M - \frac{5}{3}p$	
Division	$2p + 1$	$p^2 + 3p$	
Storage (MD volume)	$5M + 2p + 2$	$\frac{1}{2}p^2 + M + \frac{1}{2}p$	

3.2 The Algorithm of Compensation of the Movement of the RS Carrier Increasing the Accuracy of Assessment of the Degree of Meteoobjects Danger

Sounding of the atmosphere by means of an airborne RS as it was noted in the Sect. 3.2, allows to receive the assessments of spatial fields of three key MO informative parameters: RR, the average frequency and width of DS of the reflected signals.

However, DS of the signal reflected from MO and accepted by a radar station on the moving carrier depends both on characteristics of the dispersion of signals a meteoobject, and on characteristics of the relative movement of the RS carrier and MO (the Sect. 2.4.4).

The carrier of airborne RS has a considerable radial component of velocity. Besides, a flight in actual conditions is not strictly horizontal and straight. An AV during the movement makes evolutions on height, course, space angles (rolling movement and tangage) of various frequency and intensity. The specified reasons lead to emergence in the spectrum of the reflected signals of the additional components distorting the valid spectrum of radial velocities of the MO elements. In particular, the component of carrier velocity, parallel to the axis of an RS antenna system, causes change of the average frequency $\bar{f}$ of DS of the reflected signals (2.92), and the tangential component of velocity causes widening of this spectrum, that is the increase of Δf (2.93).

To exclude the influence of AV evolutions on the results of measurements, it is possible to stabilize the spatial position of an RS AS beam. Stabilization of a beam of a radar station is provided with continuous turn of the AS aperture installed on the gyrostabilized platform. It allows to compensate the accidental changes of angles of tangage and rolling motion reaching in difficult weather conditions 1° and more. According to the existing normative documents the dynamic error of the system of stabilization at the velocity of change of rolling motion 20°/s and the velocity of change of tangage 5°/s shoulf not exceed 1°. The stabilization limits for existing airborn RS of civil aviation AV are the following:

- "Groza-154": tangage—±10°, rolling motion—±15°;
- "Gradient-154": tangage—±10°, rolling motion—±40°.

However, at stabilization of the plane of the view of a radar station negative influence of components of velocity of the carrier in the horizontal plane remains.

3.2.1 The Algorithm of Compensation of the Movement of the RS Carrier with External Coherence

The compensation of influence of own movement of the RS carrier at assessment of the Doppler spectrum parameters of the reflected signals is possible if the components of the signal accepted by an RS caused by reflection from the US and MO are divided from each other on time (Fig. 3.13). In this case the signal reflected from US can be used as basic one when processing signals from MO [61, 72, 73] that corresponds to realization in a radar station of the principle of external coherence.

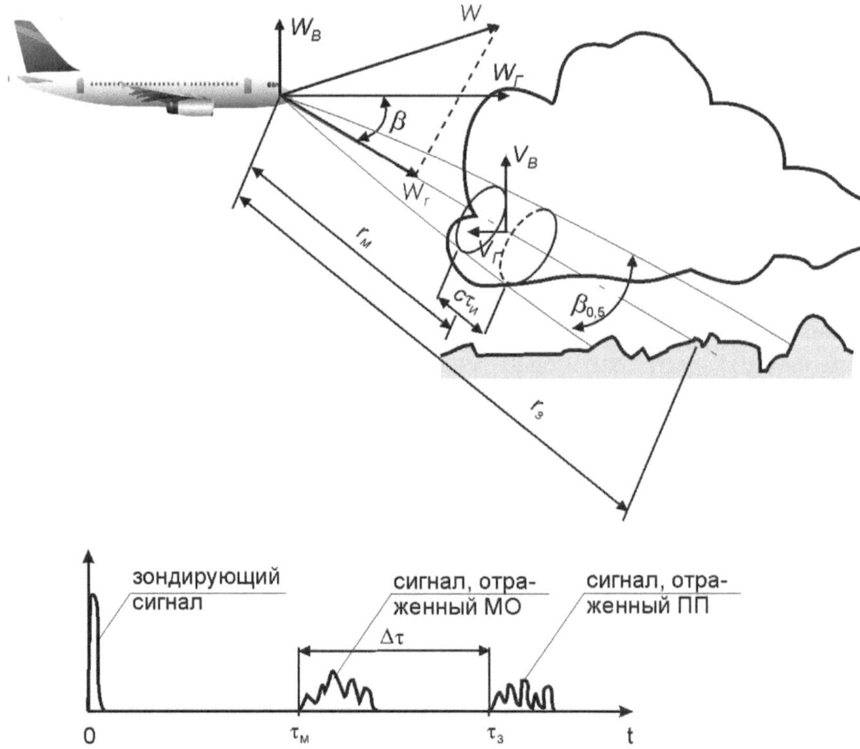

Fig. 3.13 Geometrical ratios when sounding MO and US

In the assumption of diffusion dispersion of signals of a radar station of 3-cm wave band, US can be presented in the form of sets of elementary independent reflectors, and the signal from any reflector does not surpass the sum of signals from other reflectors, that is the absence of "brilliant points" is supposed. Then each of a set of the accepted by an RS random elementary signals has the Doppler frequency proportional to the radial velocity of this reflector in relation to a radar station. As a result the value of the average Doppler frequency of the signal reflected from US is proportional to the velocity of movement of the RS carrier along reflectors on this surface

$$\bar{f}_3(t) = \frac{2f_0}{c} W_r = \frac{2f_0}{c} W \cos\theta_0 \quad (3.95)$$

Geometrical and electric characteristics of a surface influence only the relative level of signals, but do not change a physical picture of formation of the reflected signals.

For realization of the assessment (3.95) the RS carrier should be provided with the equipment of measurement of an angle of observation θ_0, and also navigation parameters (speed W of the carrier and the directions of its movement in relation a surface).

Let the signals reflected from MO and US have the form [74]

$$s_M(t) = A_M S(t - \tau_M)\cos(2\pi f_0 t + 2\pi \bar{f}_M t + \varphi_M), \quad (3.96)$$

$$s_3(t) = A_3 S(t - \tau_3)\cos(2\pi f_0 t + 2\pi \bar{f}_3 t + \varphi_3), \quad (3.97)$$

where A_M, A—the amplitude coefficients defining the reflecting properties of MO and US respectively;
S(t)—the function determined by the RS PS form;
τ_M, τ—the temporal delays of arrival of the signals reflected by MO and US, respectively equal to $\tau_M = 2r_M/c$ and $\tau = 2r/c$;
$\bar{f}_M$, $\bar{f}_3$—the Doppler frequencies caused by the relative movement of MO carrier and US carrier respectively;
φ_M, φ—the random phases at reflection of MO and US signals respectively;
r_M, r—the range to MO and US respectively.

Information on the velocity of the movement of MO in relation to US contains in the difference of instantaneous phases of the signals (3.96) and (3.97)

$$\Delta\psi(t) = \psi_M(t) - \psi_3(t) = 2\pi(\bar{f}_M - \bar{f}_3)t + (\varphi_M - \varphi_3). \quad (3.98)$$

As $\bar{f}_M = \frac{2f_0}{c}\left(W_r + \bar{V}_M\right)t$, taking into account (3.95) the expression (3.98) is transformed to the form

$$\Delta\psi(t) = \frac{4\pi f_0}{c}\left(W_r + \bar{V}_M - W_r\right)t + \left(\varphi_M - \varphi_3\right) = \frac{4\pi f_0}{c}\bar{V}_M t + \left(\varphi_M - \varphi_3\right) \quad (3.99)$$

If $\varphi_M(t)$ and $\varphi_,(t)$ are rather slow functions of time, then on the basis of (3.99) it is possible to write down the expression for the difference of instantaneous frequencies of the signals (3.96) and (3.97)

$$\Delta F = \frac{1}{2\pi}\frac{d}{dt}\Delta\psi(t) = \frac{2f_0}{c}\bar{V}_M, \quad (3.100)$$

that is the difference of frequencies ΔF does not depend on the velocity of the RS carrier.

The operation similar to (3.100) can be carried out if we detain an MO signal by the value $\Delta\tau = \tau - \tau_M$

$$s_{M3}(t) = A_M S(t - \tau_3)\cos(2\pi f_0 t + 2\pi \bar{f}_M t + \varphi_M),$$

then clip the MO (3.96) and US (3.97) signals

$$s_{M3}^K(t) = A_K(t - \tau_3)\cos(2\pi f_0 t + 2\pi \bar{f}_M t + \varphi_M),$$

$$s_3^K(t) = A_K(t - \tau_3)\cos(2\pi f_0 t + 2\pi \bar{f}_3 t + \varphi_3),$$

and multiply. Here $A_K(t - \tau)$—the signal amplitude at the output of the limiter. The total component of a signal from an output of a remultiplier is suppressed by a LFF, and differential one has the form

$$s_K^{\Delta}(t) = \frac{1}{2}A_K^2(t - \tau_3)\cos\left[2\pi f_0 t + 2\pi \bar{f}_M t + \varphi_M - (2\pi f_0 t + 2\pi \bar{f}_3 t + \varphi_3)\right]$$
$$= \frac{1}{2}A_K^2(t - \tau_3)\cos\left[\frac{4\pi f_0}{c}\bar{V}_M t + \varphi_M - \varphi_3\right].$$

Existence of a number of channels with various delays $\Delta\tau = \tau - \tau_M$, as it is noted in [73], allows to receive distribution of radial velocities of MO in relation to US. However, this approach is applicable only in case of use of the radar station with the AS having narrow DP. At increase in width of DP the efficiency of compensation sharply decreases because of the difference of Doppler frequencies of reflectors at edges of the chart leading to the essential widening of DS of the reflected signals.

3.2.2 The Algorithm of Compensation of the Movement of the RS Carrier with Internal Coherence

As the movement of the RS carrier leads to additional quasiperiodic modulation of the reflected signals, for its compensation in airborne RS with AS, not providing control of APC position (mirror AS, WSAA), the control of transceiver CO frequency (compensation on IF) or the corresponding frequency demodulation by digital methods in RSS can be used (compensation at video frequency). Higher accuracy of compensation of the movement of the carrier at insignificant complication of hardware is provided by the algorithms of signal processing on IF [62] therefore preference should be given to them.

Control of RS receiver CO frequency is provided by introduction of the relevant adjustment to it

$$f_{dv} = \frac{2f_0}{c} W \cos\alpha_0 \cos\beta_0. \tag{3.101}$$

The adjustment introduction is facilitated at using CO on the basis of a programmable frequency synthesizer. At the same time the calculator of the adjustment (3.101) can be realized in the form of the separate program module (subprogram) in memory of OBC or in the form of the separate processor device.

One of the possible algorithms of CO frequency control is based on division of CO frequency shift (an instantaneous phase) into two factors, first of which depends on the spatial provision of AS DP set by angles α_0 and β_0, the second—on the travelling speed of the carrier. Calculation in real time of the first factor synchronously changing with the DP movement demands the largest accuracy and the greatest time, however it can be replaced with matrix function which values are kept in OBC ROM that significantly reduces the time of calculation of the adjustment. The accuracy of calculation of the adjustment is defined, proceeding from the admissible shift of the central frequency of DS of the reflected signal and achievable accuracy of the assessment of own velocity of the RS carrier which values for standard coherent and pulse airborne RS should be not more than 0.05–0.10% [62]. At the same time it is necessary to consider that it is impossible to compensate the movement of the carrier ideally precisely in view of the AS DP finite sizes and impact of intra reception noise.

3.2.3 The Algorithm of "Quasimotionless RS"

As the effect of impact of the movement of the RS carrier can be reduced to change of phases of the coming signals due to change of range up to the allowed volume during repetition

$$\Delta\varphi = 2\pi f_{dv} T_n = 2\pi \frac{2W_r}{\lambda} T_n = \frac{2\pi}{\lambda} 2\Delta r, \tag{3.102}$$

for compensation of phase taper it is necessary to displace the RS ARC from the period to the repetition period by the value Δr, so that the necessary phase shift $\Delta\varphi$ was provided. Realization of this principle provides more effective compensation of the movement of the carrier, but for this purpose there should be the possibility of control of the position of APC in the longitudinal and/or cross direction in relation to vector of travelling speed of an AV providing either "immovability" of a radar station, or the movement along the direction of observation (the "quasiradial" mode).

In this case it is possible to use the algorithm of processing of signals providing "quasiimmovability" of a radar station. The essence of the specified algorithm consists in APC shift so that its position remained invariable in relation to the centre of the allowed MO volume during time of processing of the reflected signal at relocation of radar station in space [75].

As it is shown in the Sect. 2.4.4.1 the expression for a multiplier of ACF caused by the movement of the RS carrier with a PAA will have the form

$$B_{dv}(\tau) = \exp(j2\pi f_{dv}\tau)\exp\left(-\pi\left(\Delta f_y + \Delta f_z\right)\tau\right), \tag{3.103}$$

where $f_{dv} = \frac{2\Delta_r}{\lambda\tau} = \frac{2W\sec\alpha_0\sec\beta_0}{\lambda}$;

$$\Delta f_y = \frac{\Delta_y}{L_y\tau} = \frac{W\,\mathrm{tg}\alpha_0}{L_y};\quad \Delta f_z = \frac{\Delta_z}{L_z\tau} = \frac{W\,\sec\alpha_0\,\mathrm{tg}\,\beta_0}{L_z}.$$

Provision of undistorted assessment of the DS parameters of the signals reflected by MO for all temporal delays τ requires fulfilment of the condition

$$B_{dv}(\tau) = 1. \tag{3.104}$$

From (3.103) and (3.104) it follows that for the exclusion of influence of own movement of the RS carrier (at realization of a "quasimotionless" working mode of an RS) it is necessary to carry out:

- transperiod shift of APC on coordinates Y and Z by the values Δ_y and Δ_z to the point formed by crossing of the sight axis of the allowed volume, defined in initial time ($\tau = 0$) with the plane of an RS AS aperture at movement of the carrier with the velocity W during the time τ, determining the duration of process of reception of signals, undistorted by influence of own movement of the RS carrier;
- additional turn of the phase of the readings received at the shifted APC on the angle determined by the coefficient of phase correction (for the case of PAA)

$$F(\tau) = \exp\left(j\frac{4\pi}{\lambda}W\tau\sec\alpha_0\sec\beta_0\right). \tag{3.105}$$

The technical basis of realization of this algorithm can be the application as an RS AS of a phase monopulse (in two planes) antenna with weight processing of total

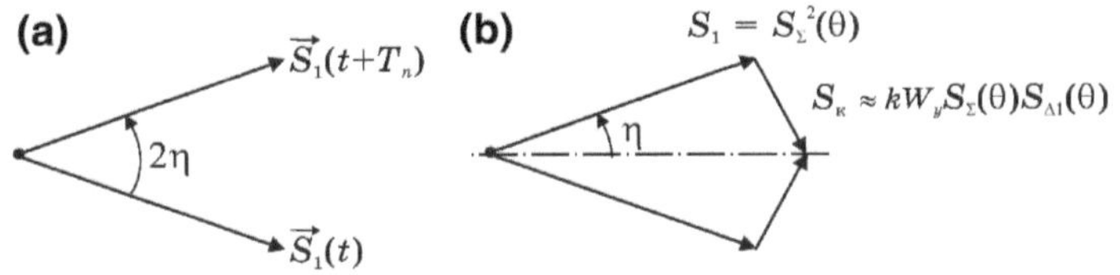

Fig. 3.14 The phase chart of the signals reflected by MO at the movement of the RS carrier

and differential signals or a planar phased antenna array (PAA) with control of the PAC position.

Let's consider the first of the specified variants [76]. For compensation of increment of the phase (3.102) in [76] it is offered to use the adjusting signal advancing the current accepted signal by π/2 and which falling behind the following accepted signal by π/2 (Fig. 3.14a). For exact compensation the equation should be correct

$$S_K(\theta) = S_1 tg\frac{\Delta\varphi}{2} = S_\Sigma(\theta) tg\frac{2\pi W T_n \sin\theta}{\lambda}, \tag{3.106}$$

where $S_\Sigma(\theta)$—the signal accepted by the total channel of the monopulse AS.

The signals $S_{\Delta 1,2}(\theta)$ of two differential channels are applied for compensation of the movement of the RS carrier of radar station along the OY and OZ axes of the BCS. Let's consider the compensation of tangential movement of the carrier along the OY axis, the result for movement along the second axis is similar.

If for transfer of the probing signal only the total channel is used, and the total channel and the relevant differential channel work for reception, then at a certain velocity the signal $S_\Sigma(\alpha)S_{\Delta 1}(\alpha)$ can play a role of the compensating signal (3.106). Here α—the azimuth of the MO sight line.

At uniform radiation by a monopulse antenna system the signals of differential and total channels are in a quadrature, and their amplitudes are connected by the ratio [76]

$$S_{\Delta 1}(\alpha) = kS_\Sigma(\alpha) tg\frac{\pi r_\Phi \sin\alpha}{\lambda}, \tag{3.107}$$

where r_F—the distance between the phase centres of two halves of the AS.

Therefore, having chosen $r_F = 2W_y T_n$, $k = 1$ and having shifted a differential signal on a phase by π/2, it is possible to receive full compensation of the movement of the RS carrier along the OY axis.

The simplified block diagram of radar station realizing this method of compensation of the RS movement is provided in the Fig. 3.15. The RS AS has three channels: total and two differential (azimuth and elevation angle) which output signals arrive to inputs of azimuth and elevation angle channels of the scheme of processing. Two signals are formed in each of the latter. The first signal is the result of addition of the corresponding total and differential signals shifted on a phase by π/2 and taken with "weight" kW_y (or kW_z).

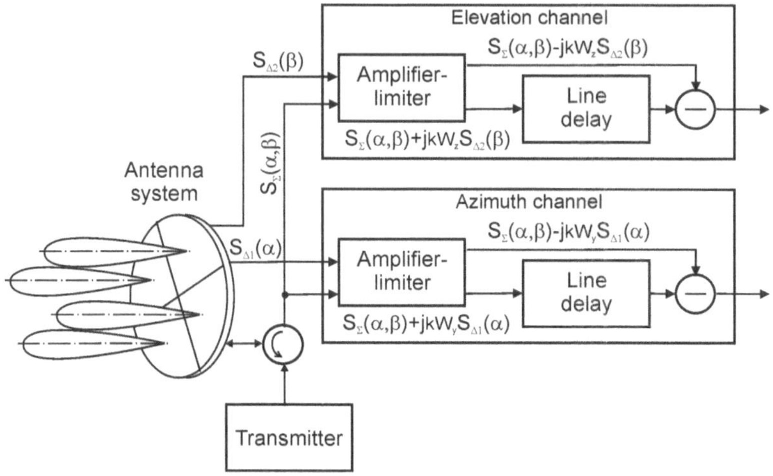

Fig. 3.15 The simplified scheme of realization of the method of APC shift for compensation of the movement of the RS carrier with the phase monopulse AS

The second signal is also the result of the addition of the corresponding total and differential signals taken with same "weight", but not shifted on a phase. Further signals are exposed to processing like single transperiod compensation in the MTS scheme. At the same time due to change of amplitude distribution of the field in an RS AS aperture occurring at addition of signals of total and differential channels the APC shift in relation to the position which it holds when this addition is not made takes place. The value of APC shift is varied by selection of "weight" kW_y of the differential channel, which is adjustment of the relation of amplitudes of signals of differential and total channels. This value is chosen depending on the velocity of flight of the RS carrier and the azimuthal provision of AS DP in the view sector.

DP of the total RS channel is chosen, proceeding from the need of provision of demanded width of its main lobe, amplification factor and LSL determined by the requirements for reliable detection of MO signals. DP of the differential channel is chosen independently, proceeding from the requirement to provide necessary ratios between the rated velocity of the movement of the carrier and admissible DP LSL.

The disadvantages of the considered variant of compensation are the need of significant phase shift of signals (3.14b) that, in turn, leads to losses in the AS amplification factor and to the DP widening and also to presence of two paths of IFA with possible dispersion of transfer coefficients. Besides, this variant of the method of compensation of the movement of the RS carrier on the basis of the APC shift is applicable only at processing of very short packs of the reflected signals [77].

The second variant of the method of compensation of the movement of the RS carrier on the basis of the APC shift can be realized in an airborne RS with flat PAA. The general block diagram of such radar station is presented in Fig. 3.16. The use of PAA allows to control amplitude-phase distribution of currents in each point of an AS aperture that is impossible at antennas with a continuous aperture.

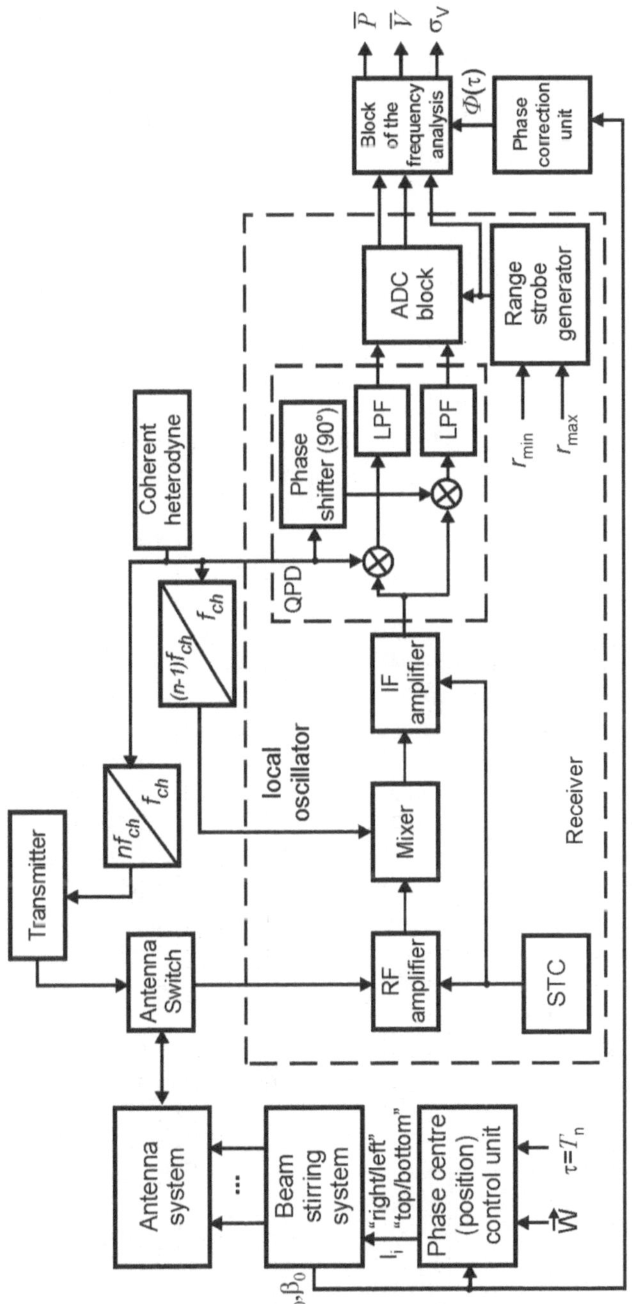

Fig. 3.16 The block diagram of the RS with PAA carrying out the assessment of the degree of MO danger with compensation of movement of the carrier

The distinctive feature of spatial characteristics is that continuous function of an aperture can be replaced with discrete function almost equivalent to it if the distance d between the elements of a discrete aperture does not exceed λ/2, where λ—the operating wavelength of radar station.

An RS includes the AS, the transceiver, the APC block, the block of the frequency analysis, the range strobe generator, the beam stirring system (BSS) of PAA, the phase centre (position) control unit (PCCU), the phase correction unit (PCU) are a part of radar station. The current values of the vector of carrier velocity $\vec{W}$ come to a radar station from other airborne systems.

The RS transceiver is constructed according to the scheme with true coherence. The transmitter generates PS which are radiated by the AS in the direction set by the BSS. The current values of azimuth α_0 and elevation angle β_0 of a maximum of AS DP from BSS come to PCCU in which the values of Δ_y and Δ_z, are calculated and also the control of PAA phase shifters is provided so that to carry out the corresponding APC shift.

The signals reflected by MO together with noise and interferences are accepted by the RS AS and come to a reception part of the transceiver where they are amplified and preliminarily filtered, and their transfer in the frequency range, convenient for processing, filtration and amplifying on IF is carried out further. The amplification factor of the processed signals is set by the scheme of sensitivity time control (STC), depending on the current range (delays of the accepted signals in relation to PS). Then the degraded signals from small ranges get to the AC line section of the RS receiver. Analog part of the receiver ends with the quadrature phase detectors (QPD) which basic signal is the CO signal.

After QPD and APC the readings of complex amplitudes of an MO signal are recorded in the RAM, and then multiplication of readings of a signal by the coefficient of phase correction (3.105) calculated by PCU is carried out. This block can have various executions, depending on the RS AS type. For example, for the case of rigidly fixed PAA the PCU block diagram realizing (3.105) is provided in Fig. 3.17.

The values of an AV air velocities, angles of course, tangage and rolling motion come to the system of compensation of the movement from the flight and navigation complex (FNC) of an AV containing the inertial navigation system (INS), the system of air signals, the attitude and heading reference system, the Doppler velocity and drift angle gauge (DVDAG).

PCCU defines the values of Δ_y and Δ_z APC shifts according to (2.97). The control of the APC position is exercised by consecutive misphasing of edge region of PAA

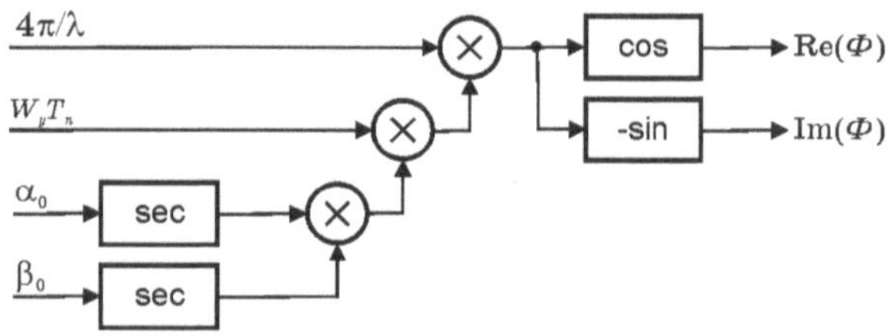

Fig. 3.17 The diagram of the block of phase correction

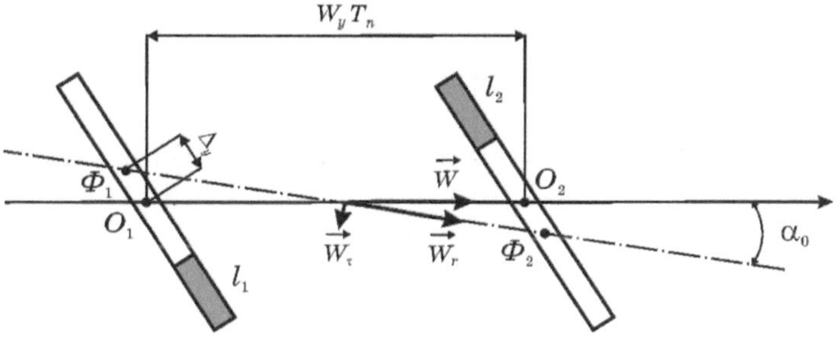

Fig. 3.18 The scheme of APC shift for compensation of the movement of the RS carrier

[62]. When misphasing the line section with the size l of the elements APC is shifted along the corresponding axis by $\Delta = \pm ld$, therefore PCCU recalculates the values of Δ_y and Δ_z into the quantity l_i of the PAA misphasing final elements (Fig. 3.18) and forms the corresponding pulse sequences arriving on registers of control of misphasing sites of PAA. Thus it is possible to provide APC shift to ±0.15 m. At the same time the changes of AS DP (widening of the main lobe, increase in LSL) practically do not depend on type of misphasing, that is on how phase shifter is installed in the disconnected region.

The output parameters of the block of the frequency analysis are the assessments of the average power of the reflected signals (RR assessments), the average radial velocity $\bar{V}$ and the RMS width of the spectrum of radial velocities ΔV for each allowed volume in the coordinates "azimuth-range". The specified assessments make the two-dimensional arrays reflecting distribution of parameters in view space.

3.2.4 The Analysis of Efficiency of the Algorithms of Compensation of the Movement of the Carrier of Coherent RS

Before the assessment of efficiency of this or that algorithm of compensation of own movement of the RS carrier it is necessary to define how it influences the accuracy of assessment of the parameters $\overline{V}$ and ΔV.

The results ($\hat{\bar{V}}$, $\Delta\hat{V}$) of R measurements of the average and RMS values of wind speed distorted by the movement of the carrier in the allowed volume of MO are defined by the expressions

$$\hat{\bar{V}}/\bar{V} \approx 1 + W/\bar{V}_M = 1 + \bar{f}_{\mathrm{dv}}/\bar{f}_M; \tag{3.108}$$

$$\frac{\overset{\wedge}{\Delta V}}{\Delta V} \approx \sqrt{1 + \left(\frac{\Delta V_{\text{dv}}}{\Delta V_M}\right)^2} = \sqrt{1 + \left(\frac{\Delta f_{\text{dv}}}{\Delta f_M}\right)^2}. \qquad (3.109)$$

The values (3.108) and (3.109) show how much the values of the corresponding assessments increase if measures for compensation of own movement of the RS carrier are not taken.

In Fig. 3.19a the dependences of the value (3.108) on the azimuth α of the allowed MO volume, calculated at $\overline{V} = 5$ m/s are presented in Fig. 3.24b, the similar dependences of the value (3.109) received at $\sigma_V = 5$ m/s are presented. It follows from the analysis of the received dependences that when using at a stage of cruiser flight of AV of civil and state aviation of airborne RS without compensation of own movement the assessment of width of the spectrum of velocities of a zone of heavy turbulence is possible with an error to 20% in the sector ±30° (at $W = 100$ m/s) and in the sector ±15° (at $W = 200$ m/s). On the other hand, in the specified sectors the influence of own movement on the assessment of $\overline{V}$ and, as a result, on efficiency of detection of dangerous WS zones is the strongest.

The efficiency of the algorithm of compensation of the movement of the carrier of airborne coherent RS can be characterized by the RMS values of errors of assessment of the parameters $\overline{V}$ and ΔV of the spectrum of MO velocities. The inaccuracy of compensation of the radial component of the carrier velocity leads to errors of the assessment of $\bar{V}$, and not ideal compensation of the tangential component—to errors of the assessment of ΔV.

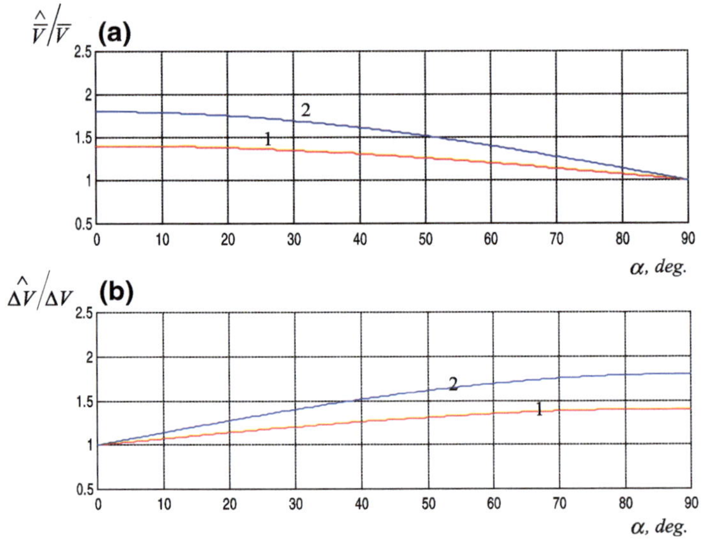

Fig. 3.19 Dependences of relative average and RMS values of wind speed on an azimuth of MO probed by airborne RS (1—W = 100 m/s; 2—W = 200 m/s)

In coherent RS the inaccuracy of compensation of the carrier velocity can be caused by two reasons:

- inexact measurement of air velocity $\vec{W}$ of the carrier and the angles (α_0, β_0), characterizing the spatial position of the axis of DP of an RS antenna system;
- inexact calculation of the relevant adjustment (3.101) and its enter in the control circuit of CO frequency.

The main source of information on an AV air velocity which is a part of FNC, at the same time air velocity is calculated by integration of these accelerometers, and as a result the dispersion of its assessment increases in proportion to duration of flight of the carrier.

The equations describing temporal dependences of errors $\vec{W}_n(t)$ of measurement of velocity of INS have the form [50]

$$\frac{\partial \vec{W}_n(t)}{\partial t} = \vec{\gamma} + \vec{\delta}, \quad \vec{W}_n(t_0) = \vec{W}_{n0},$$
$$\frac{\partial \vec{\gamma}(t)}{\partial t} = -\theta\vec{\gamma} + \sqrt{2\theta\sigma_\gamma^2}\vec{N}_\gamma, \quad \vec{\gamma}(t_0) = \vec{\gamma}_0,$$
$$\frac{\partial \vec{\delta}(t)}{\partial t} = 0, \quad \vec{\delta}(t_0) = \vec{\delta}_0,$$

- where $\vec{\gamma}$—the vector of fluctuation errors of measurement of accelerations;
- $\vec{\delta}$—the vector of the systematic and slowly changing errors of measurement of accelerations;
- $\vec{W}_{n0}$—the vector of errors of the INS initial setting on velocity representing a vector Gaussian random process with zero mathematical expectation and the standard value of the RMS deviation of 0.5 m/s;
- $\vec{N}_\gamma$—the vector of the forming Gaussian random processes;
- θ—the parameter characterizing the width of the spectrum of fluctuation errors;
- σ_γ^2—the stationary dispersion of these errors.

The standard values of the parameters of errors of INS used as a part of FNC of domestic AV are: $\theta = 100$–200 Hz, $\sigma_\gamma = (1 - 10) \cdot 10^{-2}$ m/s^2, $\delta_0 = 4.9\%$ (for the accelerometer DLUVD-5S) and 0.3–1.4% (for the accelerometer AT-1104).

In Fig. 3.20 the dependence of the error of measurement of the average velocity caused by errors of INS accelerometers and of the accumulated during processing pack of the reflected signals on duration of this time is presented. The analysis of these dependences shows when processing the pack of the reflected signals containing 20–40 impulses, the error of assessment of the average and root mean square values of wind speed can reach 10–20% (for the accelerometer AT-1104) and 40–80% (for the accelerometer DLUVD-5S).

The equipment of consumers of GLONASS and Navstar (GPS) navigation satellite systems is used for correction of slow drift of INS as a part of FNC. In particular, in the flight and navigation complex K-102 of the SU-32 plane the A-737 equipment characterized by the following measurement errors is used: 20–30 m for planned coordinates, 0.15% for velocity. In case of failure of equipment of consumers of the SNS the Sh-013 DVDAG possessing the worse, in comparison with the SNS,

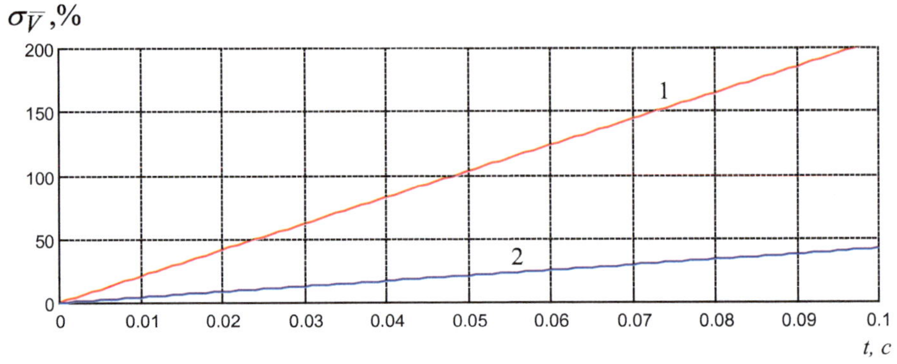

Fig. 3.20 Dependence of an error of measurement of average value of the wind speed caused by integration of errors of accelerometers on integration time duration (1—DLUVD-5S; 2—AT-1104)

precision parameters (the error of measurement of travelling speed can reach 0.25%) is used as a part of FNC [78].

The accuracy (RMS error value) of the measurements of the angles (α_0, β_0), characterizing the spatial position of the axis of DP of an RS antenna system significantly depends on the AS type (with mechanical or electronic scanning). When using PAA (Fig. 3.16) it is generally defined by the sizes of an AS aperture (the number of antenna elements), and can be accepted as equal to the width of DP in the corresponding planes which, as shown in the Sect. 3.1, should be not more than 1°. However, at the deviation of a beam of PAA from a normal to the line of location of antenna elements in the course of scanning (in the sector ±60°) there is the widening of the main lobe of DP proportional to the value $1/\cos\alpha_0$. Thereof the accuracy of determination of angular coordinates of radar station from PAA at a beam deviation from a normal significantly decreases therefore rigidly fixed PAA has the limited work area for which sizes are defined by admissible decrease in angular accuracy. So, if to consider admissible the decrease in angular accuracy twice, the maximum shift of a beam from a normal within a view zone should not exceed ±60°.

On the other hand, the accuracy of measurement of the angles (α_0, β_0), characterizing the spatial position of the axis of AS DP depends on capacity of phase shifters (PS) of antenna elements. In the majority of modern PAA four-digit (five-unit) PS are used that correspond to the step of installation 11.25° and 22.5° in the range of phase resetting 0…360°. However, as the carried-out mathematical modelling showed, in the course of compensation of the movement of the carrier due to movement of APC the changes of AS DP (widening of the main lobe, increase in level of side lobes) practically do not depend on type of misphasing, that is on how PS are installed in the disconnected region.

Partly the accuracy of the considered algorithms of compensation is influenced by stability of the frequency developed by OC of the RS receiver. However, as the level of short-term relative instability is the value of the 10^{-11} order, this circumstance can be neglected.

Besides, the efficiency of the algorithms of compensation of the movement of the carrier of an airborne RS can be characterized by the maximum volume of a pack of the reflected signals within which compensation is possible. Let's set the limits of possible values of duration of the process of reception of the undistorted reflected signals τ during the operation of radar station in the "quasimotionless" mode.

The amount M of the accepted selection of readings of the reflected signals, undistorted by the influence of the movement of the carrier, is connected with the value τ by the ratio

$$M = \mathrm{Ent}(\tau / T_n) + 1.$$

When using PAA with uniform distribution of the field in the aperture the APC shift from the point of the geometrical centre of an antenna aperture causes widening of the main lobe of DP. This widening can be quantitatively expressed by the approximate formulas

$$\varepsilon_y = \frac{\Delta\alpha' - \Delta\alpha}{\Delta\alpha} \approx \frac{L_y - \Delta_y}{L_y}, \quad \varepsilon_z = \frac{\Delta\beta' - \Delta\beta}{\Delta\beta} \approx \frac{L_z - \Delta_z}{L_z}, \tag{3.110}$$

where ε_y, ε_z—the relative widening of the main lobe of DP in the horizontal and vertical planes respectively;
$\Delta\alpha'$, $\Delta\beta'$—the width of the main lobe of DP in the horizontal and vertical planes at APC shift;
L_y, L_z—the linear sizes of an RS antenna system aperture in the horizontal and vertical planes respectively.

Taking into account (3.109) the maximal possible time of processing it is equal to

$$\tau_0 = \min\left\{\tau_{0y} = \frac{\varepsilon_y L_y}{W \sin\alpha_0}; \tau_{0z} = \frac{\varepsilon_z L_z}{W \cos\alpha_0 \sin\beta_0}\right\}.$$

The values in brackets define the greatest possible time of reception in "quasimotionless" operation mode of radar station with compensation of own movement in the horizontal (τ_{0y}) and vertical (τ_{0z}) planes.

As showed the carried-out statistical modelling, the greatest possible time of processing at $\varepsilon_y = \varepsilon_z = 0.2$ is units-tens of milliseconds at the RS carrier velocities about 100–200 m/s. As a result, during the work of radar station with the period of repetition of impulses, equal to milliseconds units, the amount of the received selection of readings of the signals reflected by MO will not exceed two-three tens.

3.2.5 The Analysis of Influence of Trajectory Irregularities and Elastic Mode of a Structure of the RS Carrier on Efficiency of Compensation of Its Movement

The influence of TI and EMS of the carrier leads to additional accidental spatial movement of APC (item 2.3) therefore APC coordinates $\left(\tilde{\Delta}_x, \tilde{\Delta}_y, \tilde{\Delta}_z\right)$ of APC estimated by the UPPC of the system of compensation of the carrier movement will differ from the necessary values $\left(\Delta_x, \Delta_y, \Delta_z\right)$, that will worsen the compensation of the movement at the solution of the problem of assessment of the parameters of a Doppler spectrum of the reflected signals. In particular, the longitudinal error will lead to decrease in accuracy of assessment of the average frequency of DS, and cross error—to decrease in accuracy of assessment of the RMS spectrum width

$$\sigma_{\bar{f}} = \frac{2(\Delta_r - \tilde{\Delta}_r)}{\lambda\tau}, \quad \sigma_{\Delta f} = \sqrt{\left[\frac{(\Delta_y - \tilde{\Delta}_y)}{L_y\tau}\right]^2 + \left[\frac{(\Delta_z - \tilde{\Delta}_z)}{L_z\tau}\right]^2}.$$

According to the thesis provided in the Sect. 2.3, the density of probability of any TI and EMS parameter at the AV movement in the turbulent atmosphere at first approximation can be approximated by normal stationary random process with zero mathematical expectation and correlation function of an exponential type (2.41). Let's present the assessment of the coefficient of phase correction (3.104) in the form

$$F(\tau) = F(\tau) + F_n(\tau), \tag{3.111}$$

where $F(\tau)$—the true value of the coefficient of phase correction;
$F_n(\tau)$—the complex stationary normal noise in the channel of phase correction caused by influence of TI and EMS, having zero mathematical expectation and dispersion σ^2_{Fn}.

Because of TI and EMS the dispersion (3.65) of the assessment of the average frequency of DS of the signals reflected by MO increases and will be

$$\sigma^2_{\bar{f}F} = \frac{(1+Q^{-1})^2(1+Q_F^{-1}) + (Q_F^{-1}-1)\rho^2(T_n)}{8\pi^2 M T_n \rho^2(T_n)}, \tag{3.112}$$

where Q—the SNR in receiving channel of a radar station;
$Q_F = \frac{1}{\sigma^2_{Fn}}$—the SNR in the channel of phase correction;
$\rho(T_n) = \exp\left[-\pi(\Delta f T_n)^2\right]$—the rated correlation coefficient.

At the high SNR in the channel of phase correction ($Q_F \to \infty$), that is at small level of TI and EMS, the expression (3.112) passes in (3.65) and $\sigma^2_{\bar{f}F} \approx \sigma^2_{\bar{f}}$. At the SNR small value ($Q_F < 1$) $\sigma^2_{\bar{f}F}/\sigma^2_{\bar{f}} \approx 1/Q_F$, however the similar situation is almost improbable. In the most probable case $Q_F > 10$ and

$$\frac{\sigma^2_{\bar{f}F}}{\sigma^2_{\bar{f}}} \approx 1 + \frac{(1 + Q^{-1})^2 + \rho^2(T_n)}{Q_F\left[(1 + Q^{-1})^2 - \rho^2(T_n)\right]}.$$

From this it follows that increase in RMS error of assessment of thc average frequency of DS in case of carrying out of operation, phase correction takes place in the conditions of observation of weak turbulence (at $\rho(T_n) \to 1$) and at small SNR in the channel of phase correction (at considerable TI and EMS).

Let us consider the influence of TI and EMS on the accuracy of assessment of RMS width of DS of the reflected signal and, respectively, on the accuracy of assessment of the spectrum of velocities of reflectors within the allowed MO volume. Let in timepoint t_1 the BCS OXYZ centre coincide with APC (Fig. 3.21). In the point M the centre of the analysed allowed MO volume is located with the coordinates (x_M, y_M, z_M). It follows from Fig. 3.21 that the azimuth and elevation angle of the centre of the allowed MO volume are consist of the values determined by the expressions

$$\alpha_M = \text{arctg}\frac{y_M}{x_M}, \quad \beta_M = \text{arctg}\frac{z_M}{x_M}.$$

In the following timepoint t_2 as a result of accidental influence of TI and EMS will move to the point O', having the coordinates $(\Delta x, \Delta y, \Delta z)$ in relation to the BCS OXYZ centre. At the same time the azimuth and elevation angle of the centre of the allowed MO volume will change

$$\alpha'_M = \text{arctg}\frac{y_M + \Delta y}{x_M + \Delta x}, \quad \beta'_M = \text{arctg}\frac{z_M + \Delta z}{x_M + \Delta x}.$$

The assessments of angles $\Delta\alpha_M = \alpha'_M - \alpha_M$ and $\Delta\beta_M = \beta'_M - \beta_M$, characterizing APC deviations from a basic trajectory owing to influence of TI and EMS, are formed by the system of the micronavigation which is a part of FNC. In the majority

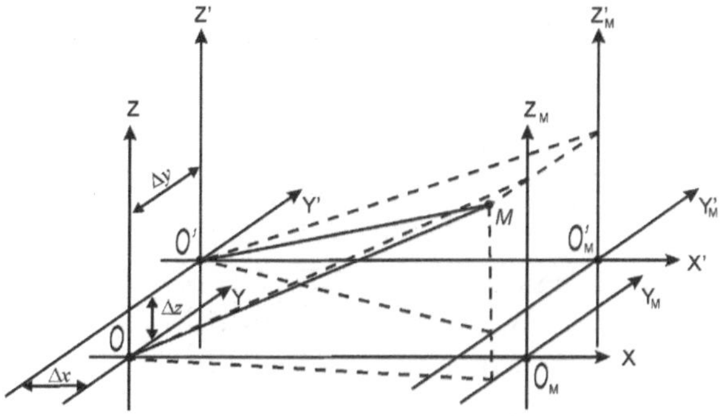

Fig. 3.21 The geometrical ratios used in the analysis of influence of TI and EMS on efficiency of compensation of the movement of the RS carrier

of practical cases these assessments can be presented in the form of the sums of mathematical expectations m_α and m_β and the aligned random components χ_α and χ_β, distributed under the normal law with the set RMS values of $\sigma_{\chi\alpha}$ and $\sigma_{\chi\beta}$. The accidental angular deviations of APC in the azimuth and elevation angle planes lead to the corresponding phase fluctuations of the reflected signal taking place in the course of accumulation of a coherent pack.

As the value of phase fluctuations is linearly connected with APC accidental side deviations and their coefficients of correlation are equal, the correlation function (CF) of phase fluctuations of the reflected signal will be described by the expression [79, 80]

$$B_\varphi(\tau) = \sigma_\varphi^2 \exp\left\{-\frac{\tau^2}{T_\varphi^2}\right\} \cos\frac{\pi\tau}{2T_{y\varphi}},$$

where

$$\sigma_\varphi^2 = \sigma_y^2 \left[\frac{4\pi}{\lambda}\right]^2 \mathrm{tg}^2\alpha_0 \tag{3.113}$$

—the dispersion of phase fluctuations;
T_φ—the interval of correlation of phase fluctuations;
$T_{y\varphi}$—the conditional interval of correlation determining the frequency of fluctuations of CF;
σ_y^2—the dispersion of side deviations of the airborne RS carrier from a trajectory.

By analogy with (3.113) the RMS error of assessment of the width of DS of the signal reflected by the allowed MO volume is connected with intensity of side movements of an AV owing to influence of TI and EMS characterized by the RMS value of velocity of these movements, taking into account (2.103) and (3.103) will have the form

$$\sigma_{\Delta f}^2 = \left(\left[\frac{2\sigma_{Wy}}{\lambda}\mathrm{tg}\,\alpha_0\right]^2 + \left[\frac{2\sigma_{Wz}}{\lambda}\sec\alpha_0 \mathrm{tg}\,\beta_0\right]^2\right), \tag{3.114}$$

where σ_{Wy}^2 and σ_{Wz}^2—the RMS values of increments of velocity of the airborne RS carrier along the OY and OZ axes respectively arising owing to influence of TI and EMS.

In Fig. 3.22a the results of calculation of dependence of the RMS error $\sigma_{\Delta f}$ of assessment of the width of DS of the reflected signal from the value of an angular deviation of the axis of DP of PAA for various values of linear velocities of TI and EMS, are presented, in Fig. 3.22b—the similar dependences of the RMS error $\sigma_{\Delta V}$ of assessment of width of the spectrum of MO velocities. It was supposed at calculation that $\beta_0 = 0$, thereof linear movements take place only in the horizontal (azimuthal) plane.

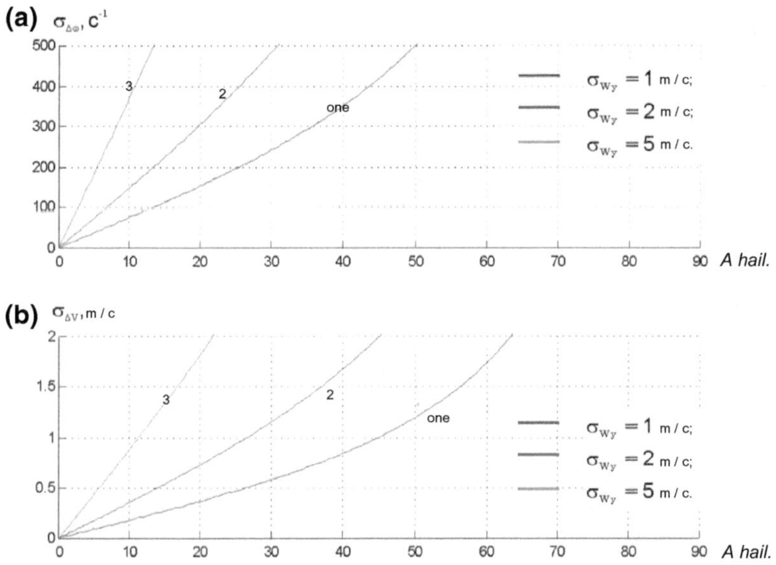

Fig. 3.22 Dependences of RMS values of errors of assessment of the width of DS of the reflected signals **a** and the width of the spectrum of MO velocities **b**, caused by the action of TI and EMS from AS DP axis azimuth (1—$\sigma_{Wy} = 1$ m/s; 2—$\sigma_{Wy} = 2$ m/s; 3—$\sigma_{Wy} = 5$ m/s)

The analysis of the results of calculation presented in Fig. 3.22 shows that at the deviation of a beam of PAA from the direction of flight of the carrier of airborne RS the error $\sigma_{\Delta V}$ of assessment of the width of the spectrum of velocities (and, respectively, an error $\sigma_{\Delta f}$ of the width of DS of the reflected signal) also significantly increases. At the same time the values of an error also depend on intensity of movements of an AV. So, at the RMS value σ_{Wy} of linear velocity of an AV in the side direction equal to 1 m/s, the error $\sigma_{\Delta V}$ does not exceed the maximum permissible value of 1 m/s (the Sect. 1.4.4) at the AS beam deviation on an azimuth in zone 0–15°, and at the RMS value of linear velocity of TI and EMS $\sigma_{\Delta V} = 5$ m/s the error $\sigma_{\Delta V}$ will not exceed 1 m/s at the AS beam deviation on an azimuth in the zone 0–12°. The quality of compensation of phase changes of the accepted R signals caused by changes of the position of PAC at accumulation of a coherent pack depends significantly on installed airborne sensors (accelerometers). These sensors should have not only high precision of measurement of accelerations, but also rather a wide band of frequencies of the measured signals which should block the frequencies of TI, EMS and aerodynamic vibrations of an AV.

Summing up the result, we will note the following:

1. The spatial stabilization of a beam of an RS AS provided by a continuous turn of the aperture installed on the gyrostabilized platform allows to compensate random changes of angles of tangage and rolling motion, however the negative influence of components of the carrier velocity in the horizontal plane remains.

2. Rather effective compensation of influence of own movement of the RS carrier at assessment of the DS parameters of the reflected signals is possible if the components of the signal accepted by an RS caused by reflection from the underlying surface (US) and MO are divided from each other on time. Then the US signal can be used as a basic one at processing of MO signals by means of an airborne RS with external coherence. However, compensation of the movement of the RS carrier on the basis of use of a basic US signal is possible only in case of use of a radar station with the antenna system having narrow DP. At increase in the DP width the efficiency of compensation sharply decreases because of difference of Doppler frequencies of reflectors at edges of DP leading to the essential widening of DS of the reflected signals.
3. For compensation of the movement of the carrier in truly coherent airborne RS with the AS, not providing the possibility of control of APC position (mirror AS, WSAA), the control of transceiver CO frequency (compensation on IF) or the corresponding frequency demodulation of the reflected signal by digital methods in RSP can be used (compensation at video frequency). Digital processing of signals on IF is more preferable in terms of efficiency of compensation. Control of CO frequency is provided by introduction of the relevant adjustment to it. The adjustment introduction is facilitated at using CO on the basis of a programmable frequency synthesizer. At the same time the calculator of the adjustment can be realized in the form of the separate program module (subprogramme) in memory of OBC or in the form of the separate processor device.
4. For compensation of the movement of the carrier in coherent airborne RS with PAA, providing possibilities of APC control, the method of transperiod control of its position in the longitudinal and/or cross direction in relation to a vector of traveling speed of an AV supporting "quasiimmovability" of a radar station can be used: as the movement of the RS carrier leads to the change of phases of the accepted signals due to the change of range up to the allowed volume during repetition, for compensation of phases taper it is enough to displace an RS APC so that its position remained invariable in relation to the centre of the allowed MO volume during the time of processing of the reflected signal. For realization of an RS quasimotionless mode it is necessary to carry out:

- a transperiod shift of APC on Y and Z coordinates (compensation of side movements of the carrier);
- a turn of a phase of readings received at the displaced APC on the angle compensating the radial movement of the carrier.

The technical basis of realization of this algorithm can be the application as an RS AS of a phase monopulse (in two planes) antenna with weight processing of total and differential signals or a planar PAA with control of the phase centre position. The disadvantages of the first variant are the need of significant phase shift of signals that, in turn, leads to big losses in the AS amplification factor and to the AS DP considerable widening and also to presence of two paths of IFA with possible dispersion of transfer coefficients. Besides, this variant of compensation is applicable only at processing of very short packs of the reflected signals.

In second variant the control of the APC position is exercised by consecutive misphasing of edge region of PAA. At the same time the changes of AS DP (widening of the main lobe, increase in level of side lobes) practically do not depend on type of misphasing, that is on how PS are installed in the disconnected region. However, the running time of a radar station in the similar mode is units-tens of milliseconds (the pack of the accepted signals does not exceed two-three tens of readings).

The influence of TI and EMS of the carrier reduces the efficiency of compensation: longitudinal movements lead to decrease in accuracy of assessment of an average frequency of a Doppler spectrum, and cross movements—to decrease in accuracy of assessment of a spectrum width. At the same time increase in RMSE of assessment of the average frequency of DS is especially notable in the conditions of observation of weak turbulence and at small SNR in the channel of phase correction (at considerable TI and EMS). The error of assessment of width of DS of the reflected signal because of influence of TI and EMS significantly increases at a deviation of a PAA beam from the direction of flight of the airborne RS carrier. The error of assessment of a Doppler spectrum width of the reflected signal because of influence of TI and EMS significantly increases at deviation of a beam of a PAA from the direction of flight of the carrier of an airborne RS.

3.3 The Algorithms of Assessment of Spatial Win Speed Fields and the Degree of Danger of the Found MO by the Results of Measurements of Parameters of a Doppler Spectrum of the Reflected Signals

3.3.1 The Algorithm of Assessment of the Danger Degree of Wind Shear Zones

For the assessment of danger of flight in the WS conditions the so-called F-factor is now used (1.4). Let us consider that the vector of an AV air velocity is directed horizontally, that is change of flight altitude happens only because of wind impact. Then taking into account (1.5) the expression (1.4) can be transformed to the form

$$F = \frac{1}{g}\left(\frac{\partial V_x}{\partial x}\dot{x} + \frac{\partial V_x}{\partial z}\dot{z} + \frac{\partial V_x}{\partial t}\right) - \frac{V_z}{W_x}.$$

The flight and navigation parameters entering the expression (1.4) come to OBC from the following onboard measuring instruments:

- the horizontal projection W_x of the vector of air velocity of an AV—from the system of air signals (SAS);
- the components $(\dot{x}, \dot{y}, \dot{z})$ of the vector of traveling speed of an AV—from platform or not platform INS, DVDAG.

Thus, for assessment of danger of the WS areas on the basis of the analysis of values of the F-factor it is necessary for each allowed volume by means of airborne RS to estimate the values of vertical speed V_z of wind, longitudinal dV_x/dx and vertical dV_x/dz wind shears, accelerations dV_x/dt, and also into predict in RCCU by the results of the current measurements the values of projections $(\dot{x}, \dot{y}, \dot{z})$ of travelling speed and the projection W_x of air velocity of an AV (Fig. 3.23). Besides, the received values of the F-factor need to be averaged at a certain distance along a flight trajectory (according to the ICAO recommendations the distance $L =$ of 1 km [81] is accepted as the scale of averaging).

According to domestic regulatory requirements the degree of danger of horizontal wind shears is estimated at the distance of 600 m (the Table 1.1).

For assessment of danger of longitudinal wind shear the values of the average radial wind speed are measured by means of radar station in two allowed volumes adjacent on range. Then the values of V_{x1} and V_{x2} of longitudinal projection of speed are estimated on them according to the chosen algorithm of the assessment of the 3-dimensional spatial field of average wind speed. In this case according to (1.2) we will have longitudinal WS

$$|\vec{v}_x| = \frac{|V_{x1} - V_{x2}|}{\delta r \cos \beta}.$$

To compare the received value to the thresholds given in Table 1.1 it is necessary to bring it to the recommended distance of 600 m

$$|\vec{v}_{x\,н}| = 600 \cdot |\vec{v}_x| = 600 \cdot \frac{|V_{x1} - V_{x2}|}{\delta r \cos \beta}. \tag{3.115}$$

For assessment of the danger of cross WS the values of average radial speed wind of are measured by means of radar station in two allowed volumes adjacent on azimuth. Then the values of V_{y1} and V_{y2} of cross projection of speed are estimated on them according to the chosen algorithm of the assessment of the 3-dimensional spatial field of average wind speed. In this case cross WS according to (1.2)

$$|\vec{v}_y| = \frac{|V_{y1} - V_{y2}|}{r\delta\alpha \cos \beta}.$$

To compare the received value to the thresholds given in Table 1.1 it is necessary to bring it to the recommended distance of 600 m

$$|\vec{v}_{y\,н}| = 600 \cdot |\vec{v}_y| = 600 \cdot \frac{|V_{y1} - V_{y2}|}{r\delta\alpha \cos \beta}. \tag{3.116}$$

The accuracy of measurement of radial speed according to the requirements formulated in the Sect. 1.3.4 is 1 m/s therefore the difference of wind speeds is measured

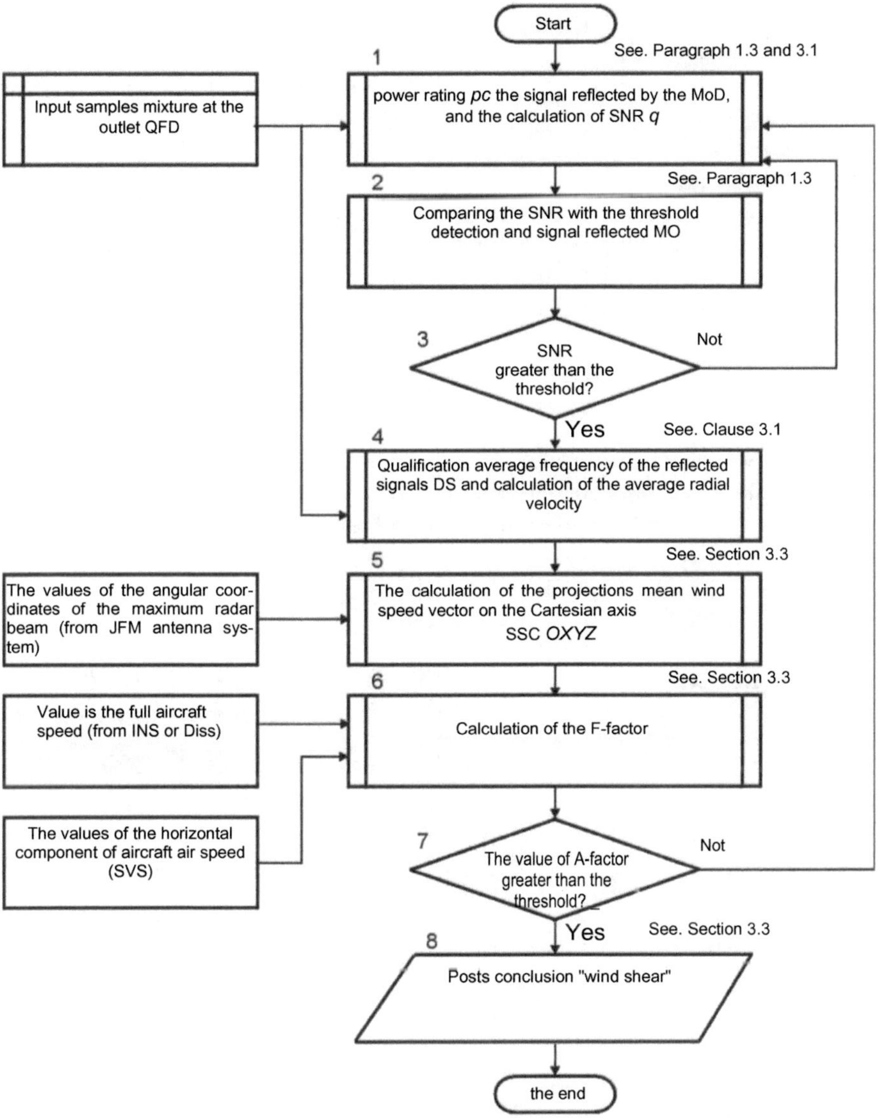

Fig. 3.23 The block diagram of the algorithm of detection and assessment of the danger of WS areas on the basis of calculation of the value of the F-factor

in the next allowed volumes with the accuracy ±of 2 m/s. It means that the first gradation of wind shear (Table 1.2) is reached already at the expense of an error of measurements, therefore, for more reliable assessment it is necessary to carry out averaging of results on several consecutive measurements.

The assessment of vertical WS is determined by a gradient of horizontal wind on height. For this purpose the values of average radial wind speed are measured by means of radar station in two allowed volumes adjacent on range. Then the values of V_{x1} and V_{x2} of longitudinal projection of speed are estimated on them according to the chosen algorithm of the assessment of the 3-dimensional field of average wind speed. In this case vertical WS is defined according to (1.2)

$$|\vec{v}_z| = \frac{|V_{x1} - V_{x2}|}{\delta h} = \frac{|V_{x1} - V_{x2}|}{\delta r \sin\beta}.$$

To compare the received value to the thresholds given in Table 1.2 it is necessary to bring it to the recommended height of 30 m

$$\left|\vec{v}_{z\,H}\right| = 30 \cdot \left|\vec{v}_z\right| = 30 \cdot \frac{\left|V_{x1} - V_{x2}\right|}{\delta r \sin\beta}. \tag{3.117}$$

For definition of qualitative characteristic of the found wind shear ("weak", …, "dangerous") the maximum of the values $\left|\vec{v}_{x\,H}\right|$ and $\left|\vec{v}_{z\,H}\right|$ is used (Fig. 3.24).

3.3.2 The Algorithm of the Assessment of the Three-Dimensional Field of Average Wind Speed

The radial component of wind speed along the sight line is defined by the expression

$$V_r = \vec{V} \times \vec{r}, \tag{3.118}$$

where $\vec{V}$—the wind speed vector;
$\vec{r} = \vec{i}\cos\alpha\cos\beta + \vec{j}\sin\alpha\cos\beta + \vec{k}\sin\beta$—the vector in which direction the speed is defined;
$\left(\vec{i}, \vec{j}, \vec{k}\right)$—OXYZ BCS unit axes.

As $\vec{V} = V_x\vec{i} + V_y\vec{j} + V_z\vec{k}$, then from (3.118) it follows that

$$V_r = V_x\cos\alpha\cos\beta + V_y\sin\alpha\cos\beta + V_z\sin\beta. \tag{3.119}$$

Thus, there is the problem of assessment of the 3-dimensional vector field (V_x, V_y, V_z) of average wind speed on the basis of the spatial field of scalar

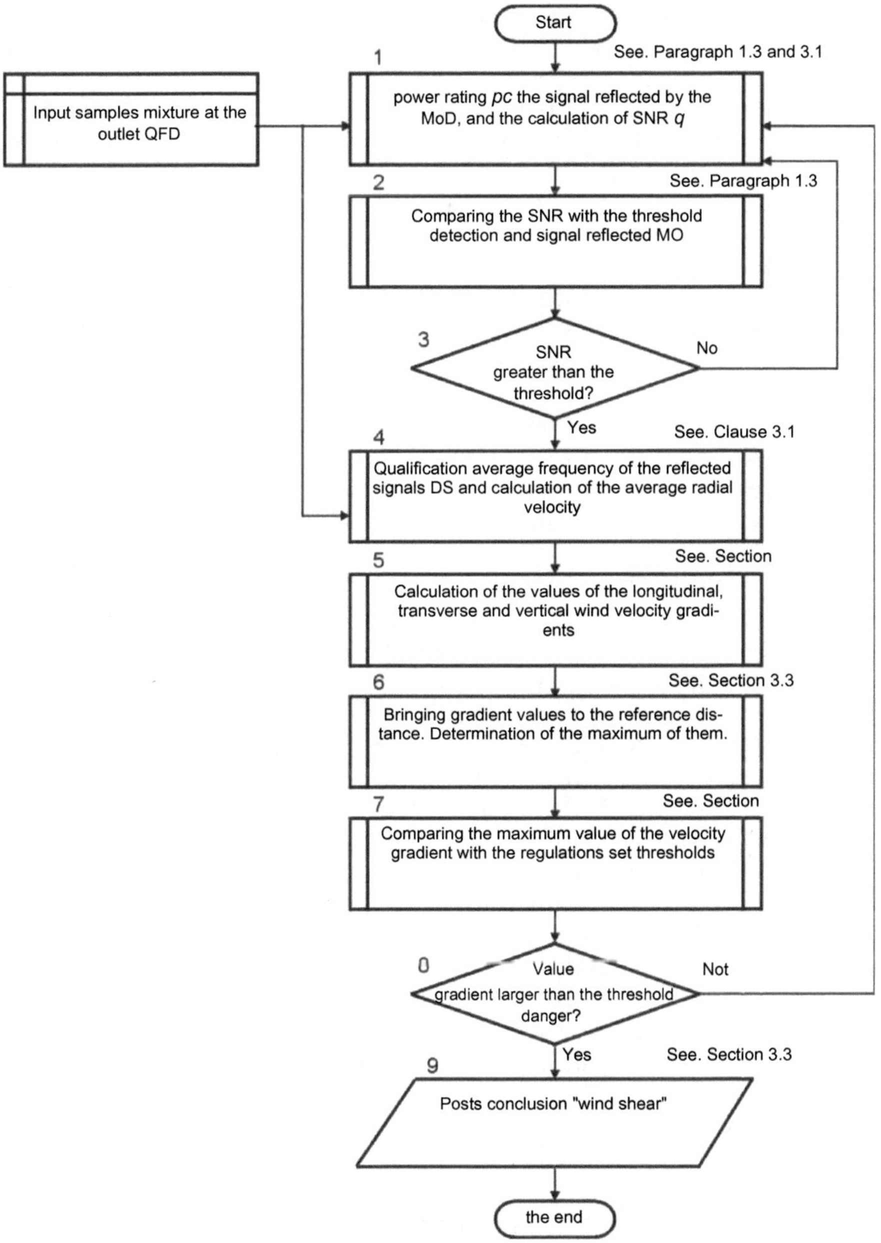

Fig. 3.24 The block diagram of the algorithm of detection and assessment of danger of areas of WS on the basis of comparison the wind speed gradient with standard established threshold values

informative parameter (average radial velocity of particles of the allowed MO volume or average frequency of DS of the reflected signals) received by means of radar station.

The development of methods of the solution of this task in relation to land meteorological Doppler radar stations began in the 1960s. The VAD (velocity azimuth display) method offered by R.M. Lhermitte) and Atlas [9, 82] became the first of them. The method consists in measurement of MO radial velocity at a certain range at conic azimuthal scanning by the AS with a constant elevation angle. If wind is spatially uniform, then in the coordinates "azimuth (time)—velocity (doppler frequency)" the radial velocity changes under the sinusoidal law which phase and amplitude are defined by the direction and the wind speed module respectively. A wind hodograph in a layer is received by means of measurements in several allowed volumes along an RS beam.

The generalized option of this method (VAD wind profile, VWP) is based on representation of the wind field of by a Fourier series which coefficients depending on the average wind speed, divergence and rotation of the wind field are determined on the basis of the analysis of distribution of radial velocity of particles on MO volume.

Another method (volume velocity processing, VVP) offered by P. Waldteufel and. H. Corbin) uses the multidimensional regression analysis of distribution of radial velocities of MO particles on volume in 3-dimensional space [9, 82]. It is in many respects similar to the VAD method considered above, but uses the function of another type for the field approximation. The accuracy of the assessment of kinematic characteristics of wind by these methods depends on a form and amount of the allowed volume, errors of measurement of radial velocity, number of parameters in model of wind, presence of not uniformity in the wind field and also real value of wind speed.

In practice of land meteorological radar-location the UWT (uniform wind technique) method based on the assumption of smoothness of the wind field and assessment of components of wind tangential to an RS beam, [82, 83] is also used. This method provides higher spatial resolution in comparison with VAD and VVP, however it is applied only at small values of an elevation angle ($\beta \leq 4°$), that, in turn, leads to additional errors because of reflections from US.

At assessment of the average wind by meteorological lidar complexes [84] the three-beam scheme of measurements which principle of action is similar to DVDAG is applied in some cases. The scheme allows to define wind speed components by results of measurement of the radial velocity in three directions (beams), however it demands additional significant completion of not only the software, but also hardware of a radar station.

On the basis of the above information the most perspective method of assessment of 3-dimensional spatial field of average wind speed at the solution by an airborne RS of the problem of detection of dangerous WS zones is the UWT method. At the sector view in a front hemisphere at small elevation angles radial velocity (3.119) can be approximately presented in the form

$$V_r \approx V_x \cos\alpha + V_y \sin\alpha. \tag{3.120}$$

The tangential component of full speed located in the horizontal plane can be defined as

$$V_{\tau} = V_x \sin\alpha - V_y \cos\alpha \quad \text{or} \quad V_{\tau} = -\partial V_r/\partial\alpha. \tag{3.121}$$

Beside (3.119) and (3.120), the fundamental equation of continuity of the wind field [58] can be used for search of the components (V_x, V_y, V_z) of full wind speed [58]

$$\frac{d\rho}{dt} + \rho \operatorname{div}\vec{V} = 0,$$

where ρ—air density. In the assumption of incompressibility of air the equation of continuity will be transformed to the form

$$\operatorname{div}\vec{V} = 0$$

or in the Cartesian coordinates

$$\frac{\partial V_x}{\partial x} + \frac{\partial V_y}{\partial y} + \frac{\partial V_z}{\partial z} = 0. \tag{3.122}$$

Uniting (3.120), (3.121) and (3.122) in one system, we receive

$$\begin{cases} V_x \cos\alpha + V_y \sin\alpha = V_r, \\ V_x \sin\alpha - V_y \cos\alpha = -\partial V_r/\partial\alpha, \\ \frac{\partial V_x}{\partial x} + \frac{\partial V_y}{\partial y} + \frac{\partial V_z}{\partial z} = 0. \end{cases} \tag{3.123}$$

Solving the system (3.123), we receive the following expressions for the component of full wind speed

$$\begin{cases} V_x = V_r \cos\alpha - \frac{\partial V_r}{\partial\alpha} \sin\alpha, \\ V_y = V_r \sin\alpha - \frac{\partial V_r}{\partial\alpha} \cos\alpha, \quad (3.124) \\ V_z = \int\limits_{-z_0}^{z_1} \left(\frac{\partial V_x}{\partial x} + \frac{\partial V_y}{\partial y}\right) dz, \quad (3.125) \end{cases}$$

that is for calculation of a vertical component of wind speed in this point at height z_1 it is necessary to know gradients of longitudinal and cross horizontal components of speed at heights from $-z_0$ to z_1. Here z_0—is the AV flight altitude.

On the other hand, when using the UWT method within the assumptions made above the component V_z of full wind speed can be found in the Cartesian BCS from geometrical ratios. For projections to the OZ the following expression is correct

$$V_z = V_r \sin\beta - V_{\tau\text{в}} \cos\beta\,, \tag{3.126}$$

where $V_{\tau\text{в}}$—the tangential component of full wind speed of in the vertical plane.

Then we will consider the module of the vector of full wind speed

$$|V|^2 = V_x^2 + V_y^2 + V_z^2 = V_r^2 + V_{\tau\text{г}}^2 + V_{\tau\text{в}}^2\,.$$

Substituting in this equation the value $V_{\tau\text{в}}$ from (3.126) and grouping members with identical order, we will receive the quadratic equation of the form

$$V_z^2 \sin^2\beta - 2(V_r \sin\beta)V_z + \left(V_r^2 + \left[V_{\tau\text{г}}^2 - V_x^2 - V_y^2\right]\cos^2\beta\right) = 0\,,$$

in which we will substitute the values $V_{\tau\text{г}}$ (3.121), V_x (3.124) and V_y (3.125), then

$$V_z^2 \sin^2\beta - 2(V_r \sin\beta)V_z + \left(V_r^2 \sin^2\beta + 2V_r \frac{dV_r}{d\alpha} \sin 2\alpha \cos^2\beta\right) = 0.$$

From two solutions of this quadratic equation

$$V_{z\,1,2} = \frac{V_r \pm \cos\beta \sqrt{V_r^2 - 2V_r \frac{dV_r}{d\alpha} \sin 2\alpha}}{\sin\beta}$$

we choose the value smaller on the module

$$V_z = \frac{V_r - \cos\beta \sqrt{V_r^2 - 2V_r \frac{dV_r}{d\alpha} \sin 2\alpha}}{\sin\beta}. \tag{3.127}$$

Thus, the values of projections (V_x, V_y, V_z) of the 3-dimensional vector field of average wind speed to axes of the Cartesian BCS taking into account the introduced restrictions can be estimated, according to (3.124), (3.125) and (3.127), by the analysis of the spatial field of scalar informative parameter (average radial speed of particles of the allowed MO volume or average frequency of DS of the reflected signals). The accuracy of the corresponding assessments is defined, generally by the accuracy of measurement of parameters (V_r, α, β) by means of airborne RS.

3.3.3 *The Algorithm of the Assessment Danger of Zones of the Increased Atmospheric Turbulence*

The degree of dangerous impact of atmospheric turbulence on an AV according to the ICAO recommendations is defined by the overload n, which is functionally connected with the velocity ε of dissipation of turbulent kinetic energy (Table 1.1). On the other hand, the values ε influences the turbulent contribution to the width ΔV of the spectrum of HM velocities within the allowed volume

$$\varepsilon = \begin{cases} \frac{1{,}3\Delta V_t^3}{a(1-\gamma_1/15)^{2/3/3}}, & \text{when } a \geq b; \\ \frac{1{,}3\Delta V_t^3}{b(1-\gamma_2/15)^{2/3}}, & \text{when } a < b, \end{cases} \quad (3.128)$$

where $a = r\delta\alpha$ and $b = c\tau_u/2$—are the cross and longitudinal sizes of the allowed volume of radar station;

$$\gamma_1 = 1 - b^2/a^2; \quad \gamma_2 = 1 - a^2/b^2.$$

The assessment of the width ΔV of the spectrum of HM velocities by results of R measurements is proportional to the width of DS of the signals reflected by the allowed MO volume, is defined as

$$\Delta V = \Delta f\lambda/2.$$

However, as shown in the Sect. 3.1.3.2, the value ΔV, besides ΔV_t, contains a number of the additional components distorting a real form of the spectrum of HM velocities,

$$\Delta V^2 = \Delta V_t^2 + \Delta V_{ws}^2 + \Delta V_g^2 + \Delta V_{sc}^2,$$

where $\Delta V_{ws} = \frac{|\vec{v}_\perp| r \Delta\beta}{2\sqrt{2\ln 2}} \cos\beta$, $\Delta V_g = \Delta V_g^0 \sin\beta$, $\Delta V_{sc} = \frac{\Phi_\alpha \lambda}{4\delta\alpha T_v}$—the components of the width of the spectrum of velocities caused by influence of WS, gravitational falling of HM, scanning of DP of an antenna system of radar station; $\Delta V_g^0 = 0{,}21Z^{0{,}08}$.

Thus, the algorithm of assessment of danger of areas of atmospheric turbulence (Fig. 3.25) consists in calculation of width of the spectrum of velocities ΔV, allocation from it of the component ΔV_t, caused by the influence of turbulence, calculation of the value ε and its comparison with the threshold values given in the Table 1.1.

Summing up the result, it is possible to say the following:

1. The degree of danger of WS can be estimated either on the predicted value of the F-factor characterizing the change of total energy of an AV owing to influence of SW or by comparison of the SW values brought to a reference distance with standard established thresholds. The first approach is more informative, however it demands continuous measurement and forecasting of a large number of

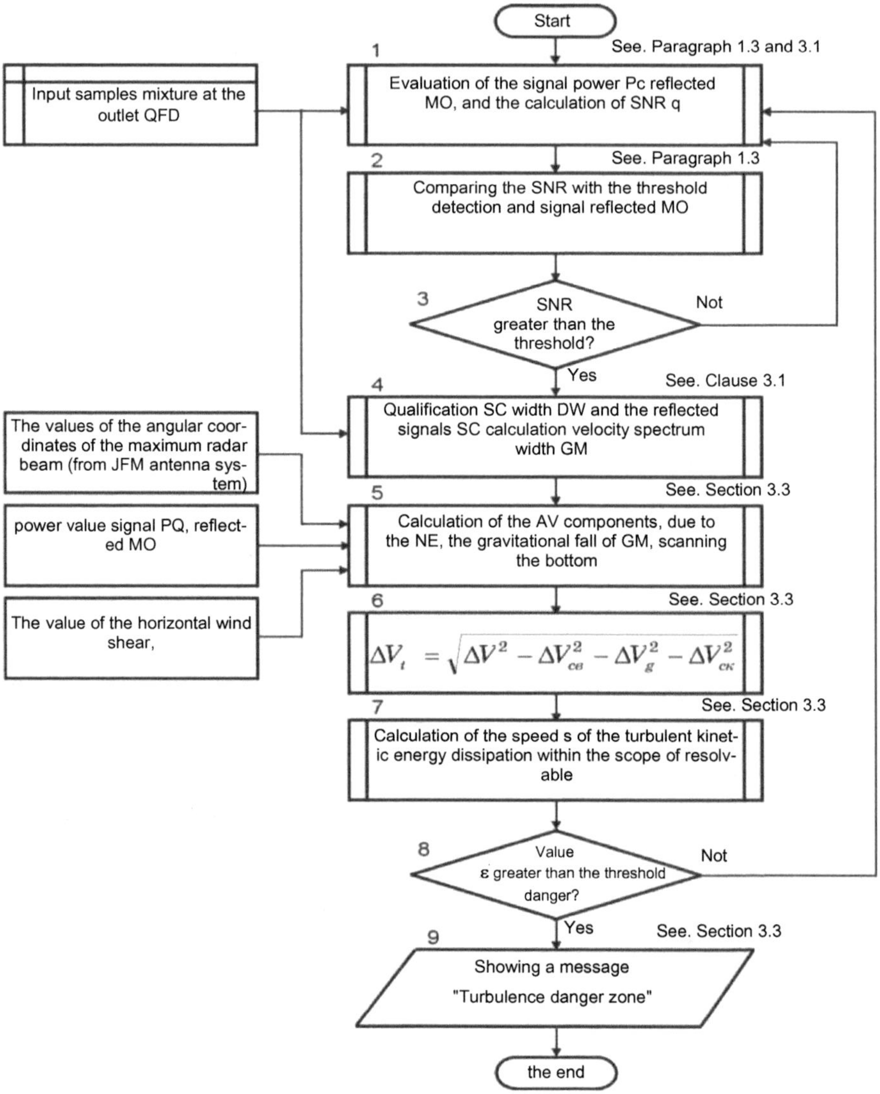

Fig. 3.25 The block diagram of the algorithm of detection and assessment of danger of areas of atmospheric turbulence on the basis of comparison of values of velocity of dissipation of its kinetic energy with standard established threshold values

kinematic parameters of the AV movement. Besides, this method does not consider shears of cross and vertical wind components.

2. For adequate assessment of any indicator of danger of WS it is necessary to estimate the 3-dimensional spatial field of average wind speed. The accuracy of performance of this operation depends on the accuracy of measurement of average frequency of DS of the reflected signals (that is the average radial velocity of MO) and also on the accuracy of measurement of an angular position of the main lobe of DP of an RS antenna system.
3. The VAD, VWP, VVP, UWT methods and also a number of their modifications are developed for the solution of the specified task in relation to land meteo RS. The accuracy of assessment of characteristics of wind by these methods depends on a form and amount of the allowed volume, errors of measurement of radial velocity, number of parameters in model of wind, presence of not uniformity in the wind field and also real value of wind speed.
   The most perspective of the specified methods for realization in airborne RS is the UWT method as it provides higher spatial resolution in comparison with other methods, however it is applicable only at small values of an elevation angle ($\beta \leq 4°$), that, in its turn, leads to additional errors because of reflections from US.
4. The degree of dangerous impact of atmospheric turbulence on an AV is defined by the value ε of velocity of dissipation of its energy which is functionally connected with a turbulent contribution to the width ΔV of the spectrum of HM velocities. As a result the algorithm of assessment of danger of areas of atmospheric turbulence consists in calculation of width of the spectrum of velocities ΔV, allocation from it of the component ΔV_t, caused by the influence of turbulence, calculation of the value ε and its comparison with standard established threshold values.

The main conclusions for the Sect. 3.3

On the basis of the results of the third section it is possible to make the following main conclusions:

1. The algorithms of assessment of the DS parameters of MO signals at their realization in airborne RS should meet the following requirements:

   – ensuring work in the conditions of aprioristic uncertainty about parameters of the processed signals;
   – ensuring work in the time scale close to real;
   – the simple (single-channel) realization suitable for airborne RS with digital processing.

2. Realization of the ML method in the conditions of spatial heterogeneity of the estimated parameters demands carrying out of large volume of calculations and considerable time for receiving of necessary assessments for each allowed volume of RS. The nonparametric methods of assessment of the moments of DS of MO signals which are widely applied in land meteo radar stations (the method

of paired pulses and periodogram method) demand the significantly smaller volume of calculations and memory. However, the accuracy and resolution of assessments provided by them are limited to the value, inverse to duration of selection (pack) of the accepted signals that it is especially essential to airborne RS of modern high-speed AV when the time of R contact with MO is extremely limited that does not allow to receive the selection of large volume.

3. In similar conditions the block parametric methods of assessment of AR coefficients can be applied for processing of packs of the reflected signal of some fixed volume. The most preferable among them is the modified covariant method based on joint minimization of RMS errors of linear prediction forward and back and having the largest accuracy among AR methods.
4. The methods based on the analysis of the own values of ACM of the accepted signals (MUSIC, ESPRIT, the minimum norm, etc.), have the accuracy of frequency estimates, comparable with the covariant method however the focus of the specified methods on allocation of R of signals of the targets against the background of broadband noise complicates their use at the solution of the problem of assessment of the moments of DS of such spatially distributed target as MO. In particular, the methods based on the analysis of the own values of ACM lead to wrong (ambiguous) measurements in all cases when the component DS of MO signals are strongly correlated, and the accuracy of assessment of the DS parameters by the MUSIC method with widening of a spectrum worsens much quicker, than for the covariant method. At the solution of the problem of detection of the zones of intensive atmospheric turbulence forming the reflected signals with wide DS, preference should be given to the modified covariant method. This conclusion is especially relevant at installation of airborne RS on the high-speed carrier.
5. The main disadvantages of the majority of recursive methods are the large volume of the calculations carried out on each step of a recursion in real time and also small velocity of convergence of the received assessments to true values of parameters, which significantly complicate their use when processing the signals reflected by MO in airborne RS of modern high-speed AV when time of R contact with the target is extremely limited.
6. The accuracy of assessment of the moments of DS of the reflected signals, besides the factors determined by the used processing method is influenced by physical characteristics of the studied MO (presence of longitudinal and cross WS, gravitational falling of HM of the different size by gravity) and technical parameters of airborne RS (width and velocity of scanning of DP by AS, stability of the carrier frequency of PS of radar station, bandwidth of RS receiver, impulse duration, etc.). However, the influence of the majority of these factors is either insignificant or is shown only at great values of an elevation angle and wide DP.
7. The Berg method is the steadiest among parametric methods of spectral assessment. Realization of the covariant method on the basis of arithmetics of usual accuracy (32 categories) leads to relative errors of assessment of the DS parameters of MO signals equal to units and even tens of percent. The reason consists that these algorithms are based on matrix algebra which is very sensitive to

the capacity of the processed data. When using the arithmetics of the doubled accuracy (64 bits) the relative error of covariant algorithms obviously is less than one percent, which is sufficient for practical purposes.

8. The spatial stabilization of a beam of an RS AS due to continuous turn of the aperture installed on the gyrostabilized platform allows to compensate random changes of angles of tangage and rolling motion, however does not provide elimination of negative influence of components of the carrier velocity in the horizontal plane.
9. In modern and perspective coherent airborne RS the control of the frequency of the coherent oscillator (CO) of the transceiver (compensation at intermediate frequency) or the corresponding frequency demodulation of the reflected signal by digital methods in RSP can be used (compensation at a video frequency) for compensation of radial relocation of the carrier movement. The transperiod shift of RS APC can be used for compensation of tangential movement of the carrier so that its position remained invariable in relation to the centre of the allowed MO volume during the time of processing of the reflected signal.
   The technical basis of realization of this method can be the application as an RS AS of a phase monopulse antenna with weight processing of total and differential signals or a planar PAA. The disadvantages of realization of the variant with the monopulse antenna are the need of significant phase shift of signals that, in turn, leads to big losses in the AS amplification factor and to the AS DP considerable widening and also to presence of two paths of IFA with possible dispersion of transfer coefficients. Besides, this variant is applicable only at processing of short packs of the reflected signals. In airborne RS with a planar PAA the control of the APC position is exercised by consecutive misphasing of edge region of PAA. At the same time the changes of DP (widening of the main lobe, increase in LSL) practically do not depend on type of misphasing, that is on how PS are installed in the disconnected region. However, the running time of a radar station in the similar mode is units-tens of milliseconds (the pack of the accepted signals does not exceed two-three tens of readings).
10. The influence of TI and EMS of the carrier leads to additional random spatial movement of APC that reduces the efficiency of compensation. At the same time increase in RMS error of assessment of average frequency of DS is especially notable in the conditions of observation of weak turbulence and at small SNR in the channel of phase correction (at considerable TI and EMS). The error of assessment of width of DS of the reflected signal because of influence of TI and EMS significantly increases at a deviation of a PAA beam from the direction of flight of the airborne RS carrier, and its values are also proportional to intensity of movements of an AV.
    The quality of compensation of trajectory distortions of the accepted R signals caused by changes of the position of PAC at accumulation of a coherent pack fully depends on installed airborne sensors (accelerometers). These sensors should have not only high precision of measurement of accelerations, but also rather wide band of frequencies of the measured signals which should block the frequencies of TI, EMS and aerodynamic vibrations of an AV.

11. The degree of danger of wind shear can be estimated either on the predicted value of the F-factor characterizing the change of total energy of an AV owing to influence of SW or by comparison of the SW values brought to a reference distance with standard established thresholds. The first approach is more informative, however it demands continuous measurement and forecasting of a large number of kinematic parameters of the AV movement. Besides, this method does not consider shears of cross and vertical wind components.
12. For adequate assessment of any indicator of danger of wind shear it is necessary to estimate the 3-dimensional spatial field of average wind speed. The accuracy of performance of this operation depends on the accuracy of measurement of average frequency of DS of the reflected signals (that is the average radial MO velocity) and also on the accuracy of measurement of an angular position of the main lobe of the AS DP.
The VAD, VWP, VVP, UWT methods and also a number of their modifications are developed for the solution of the specified task in relation to land meteo RS. The accuracy of assessment of characteristics of wind by these methods depends on a form and amount of the allowed volume, errors of measurement of radial velocity, number of parameters in model of wind, presence of not uniformity in the wind field and also real value of wind speed. The most perspective method of assessment of the three-dimensional field of average speed wind at the solution by an airborne RS of the problem of detection of dangerous zones of wind shear is the UWT method, as it provides higher spatial resolution in comparison with other methods, however it is applicable only at small values of an elevation angle ($\beta \leq 4°$), that, in its turn, leads to additional errors because of reflections from US.
13. The degree of dangerous impact of atmospheric turbulence on an AV is defined by the value ε of velocity of dissipation of its energy which is functionally connected with a turbulent contribution to the width σ_V of a spectrum of velocities of hydrometeors within the allowed volume. As a result the algorithm of assessment of danger of areas of atmospheric turbulence consists in calculation of width of the spectrum of velocities ΔV, allocation from it of the component ΔV_t, caused by the influence of turbulence, calculation of the value ε and its comparison with standard established threshold values.

References

1. Sarychev V. A., Antsev G. V. Operating modes of civil multipurpose airborne RS. Radio electronics and communication, 1992, No. 4, pages 3–8
2. Gong S., Rao D., Arun K. Spectral analysis: from usual methods to methods with high resolution. - In the book: Superbig integrated circuits and modern processing of signals/Eds. Gong S., Whitehouse H., Kaylat T.; the translation from English eds. V.A. Leksachenko. - M.: Radio and Communication, 1989. - P.45–64
3. Marple Jr. S.L. Digital spectral analysis and its applications: The translation from English - M.: Mir, 1990. - 584 pages

4. Korostelev A.A. Space-time theory of radio systems: Text book. - M.: Radio and Communication, 1987. - 320 pages
5. Sloka V.K. Issues of processing of radar signals. - M.: Sov. radio, 1970. - 256 pages
6. Shirman Ya.D., Manzhos V.N. The theory and technology of processing of radar information against the background of interferences. - M.: Radio and Communication, 1981. - 416 pages
7. Chernyshov E.E., Mikhaylutsa K.T., Vereshchagin A.V. Comparative analysis of radar methods of assessment of spectral characteristics of moisture targets: The report on the XVII All-Russian symposium "Radar research of environments" (20–22.04.1999). - In the book: Works of the XVI-XIX All-Russian symposiums "Radar research of environments". Issue 2. - SPb.: VIKU, 2002 - p. 228–239
8. Melnikov V.M. Characteristics of the wind field in clouds according to the incoherent radar. News of higher education institutions: Radiophysics, 1990, v.33, No. 2, pages 164–169
9. Doviak R., Zrnich. Doppler radars and meteorological observations: The translation from English - L.: Hydrometeoizdat, 1988. - 511 pages
10. Dorozhkin N.S., Zhukov V.Yu., Melnikov V.M. Doppler channels for the MRL-5 radar. Meteorology and hydrology, 1993, No. 4, pages 108–112
11. Ryzhkov A.V. Characteristics of meteorological radar stations. Foreign radio electronics, 1993, No. 4, pages 29–34
12. Melnikov V.M. Meteorological informational content of doppler radars. Works of the All-Russian symposium "Radar researches of environments". Issue 1. - SPb.: MSA named after A.F. Mozhaisky, 1997. - p. 165–172
13. Jenkins G., Watts D. Spectral analysis and its applications: The translation from English in 2 v. - M.: Mir, 1971–1972
14. Vlasenko V. A., Lappa Yu.M., Yaroslavsky L.P. Methods of synthesis of fast algorithms of convolution and spectral analysis of signals. - M.: Science, 1990 - 180 pages
15. Alekseev V.G. About nonparametric assessments of spectral density. Radiotechnics and electronics, 2000, v. 45, No. 2, pages 185–190
16. Anderson T. Statistical analysis of time sequences. - M.: Mir, 1976
17. Zubkov B.V., Minayev E.R. Bases of safety of flights. - M.: Transport, 1987. - 143 pages
18. Mikhaylutsa K.T., Ushakov V.N., Chernyshov E.E. Processors of signals of aerospace radio systems. - SPb.: JSC Radioavionika, 1997. - 207 pages
19. Analog and digital filters: Ed. book/S.S. Alekseenko, A.V. Vereshchagin, Yu.V. Ivanov, O.V. Sveshnikov; Eds. Yu.V. Ivanov. - SPb.: BGTU VOENMECH, 1997 - 140 pages
20. Voyevodin V.V., Kuznetsov Yu.A. Matrixes and calculations. – M.: Science, 1984 – 320 pages
21. Zhuravlev A.K., Lukoshkin A.P., Poddubny S.S. Processing of signals in adaptive antenna arrays. - L.: LSU publishing house, 1983. - 240 pages
22. Ortega J. Introduction to parallel and vector methods of the solution of linear systems: The translation from English - M.: Mir, 1991. - 376 pages
23. Vereshchagin A.V., Mikhaylutsa K.T., Chernyshov E.E. Features of detection and assessment of characteristics of turbulent meteoformations by airborne doppler weather radars: The report on the XVIII All-Russian symposium "A radar research of environments" (18–20.04.2000). - In the book: Works of the XVI-XIX All-Russian symposiums "Radar research of environments". Issue 2. - SPb.: VIKU, 2002 - p. 240–249
24. Mao Yu-Hai, Lin Daimao, Li Wu-gao. An adaptive MTI based on maximum entropy spectrum estimation principle. Alta Frequenza, 1986, vol. LV, №2, p.p. 103–108
25. Mironov M.A. Assessment of parameters of the model of autoregression and moving average on experimental data. Radio engineering, 2001, No. 10, pages 8–12
26. Nonlinear Methods of Spectral Analysis, 2ed ed., S. Haykin ed., Springer-Verlag, New York, 1983
27. The algorithms of estimation of angular coordinates of sources of radiations based on the methods of the spectral analysis. Drogalin V.V., Merkulov V.I., Rodzivilov V. A. et al. Foreign radio electronics. Achievements of modern radio electronics, 1998, No. 2, pages 3–17
28. Ways and algorithms of anti-jam of airborne RS from multipoint non-stationary interferences. Drogalin V.V., Kazakov V. D., Kanashchenkov A.I., et al. Foreign radio electronics. Achievements of modern radio electronics, 2001, No. 2, pages 3–51

29. Nemov A.V. Spectral estimation with high resolution on not equidistant selection of data. News of higher education institutions. Electronics, 2001, No. 4, pages 101–108
30. Kumarsen R., Tafts D.W. Estimation of the angles of arrival of multiple plane waves. IEEE Transaction on Aerospace and Electronic Systems, 1983, vol. AES-19, January, p.p. 134–139
31. Nemov A.V., Dobyrn V.V., Kuznetsov E.V. Joint use of the superresoluting frequency assessments. News of higher education institutions of Russia. Radio electronics, 2002, No. 2, pages 85–92
32. Vereshchagin A.V., Ivanov Yu.V., Perelomov V.N., Myasnikov S. A., Sinitsyn V. A., Sinitsyn E.A. Processing of radar signals of airborne coherent and pulse radar stations of planes in difficult meteoconditions. St.-Petersburg, pub. house. Research Centre ART, 2016. - 239 pages
33. Melnichuk Yu.V., Chernikov A.A. Operational method of detection of turbulence in clouds and rainfall. Works of the CAO, 1973, issue 110, p. 3–11
34. Melnikov V.M. About definition of a spectrum of a meteoradio echo by means of measurement of frequency of emissions of an output signal of the radar. Works of VGI, 1982, issue 51, p. 17–29
35. Tikhonov V.I., Harisov V.N. Statistical analysis and synthesis of radio engineering devices and systems. - M.: Radio and Communication, 1991. - 608 pages
36. Tikhonov V.I., Kulman N.K. Nonlinear filtration and quasicoherent reception of signals. - M.: "Sov. Radio", 1975. - 704 pages
37. Sage E., Mells J. The theory of assessment and its application in communication and management. – M.: Communication, 1976. – 496 pages
38. Wax M., Kailath T. Detection of Signals by Information Theoretic Criteria. IEEE Transactions on Acoustics, Speech and Signal Processing, 1985, vol.33, №4, p.p. 387–392
39. Stoica P., Nehorai A. MUSIC, Maximum Likelihood and Cramer-Rao Bound: Further Results and Comparisons. IEEE Transactions on Acoustics, Speech and Signal Processing, 1990, vol.38, №12, p.p. 2140–2150
40. Xu Xioa-Liang, Bucklej K.M. Bias and variance of direction of arrival estimates from MUSIC, MIN-NORM and FINE. IEEE Transaction on Signal Processing, 1994, vol. 42, [1]7, p.p. 1181–1186
41. Srinivas K.P., Reddy V.U. Finite data performance of MUSIC and minimum norm methods. IEEE Transaction on Aerospace and Electronic Systems, 1994, vol. AES-30, №1, p.p. 161–174
42. Abramovich Yu.I., Spencer N.K., Gorokhov A.Yu. Allocation of independent sources of radiation in not equidistant antenna arrays. Foreign radio electronics. Achievements of modern radio electronics, 2001, No. 12, page 3–17
43. Proukakis C., Manikas A. Study of ambiguities of linear arrays. - In the book: Proc. ICASSP-94, Adelaide, 1994, vol.4, p.p. 549–552
44. Kravchenko N.I., Bakumov V.N. Limit error of measurement of regular doppler shift of frequency of meteorological signals. News of higher education institutions. Radio electronics, 1999, No. 4, pages 3–10
45. Kulikov E.I. Extreme accuracy of measurement of the central frequency of narrow-band normal random process against the background of white noise. Radio engineering and electronics, 1964, No. 10, pages 1740–1744
46. Repin V.G., Tartakovsky G.P. Statistical synthesis at aprioristic uncertainty and adaptation of information systems. - M.: "Soviet Radio", 1977. - 432 pages
47. Zrnic D. Estimation of Spectral Moments for Weather Echoes. IEEE Transactions on Geoscience, 1979, v. GE-17, [1] 4, p. 113–128
48. Vereshchaka A.I., Olyanyuk P.V. Aviation radio equipment: The textbook for higher education institutions. - M.: Transport, 1996. - 344 pages
49. Current state and some prospects of development of radar station of fighters: Abstract NZNT, 1987, No. 14, pages 11–17
50. Vereshchagin A.V., Mikhaylutsa K.T., Chernyshov E.E. Ways of improvement of processing of signals in airborne meteo radar stations. In the book: Modern technologies of information extraction and processing: Collection of scientific works. – SPb.: JSC Radioavionika, 2001. – p. 211–230

51. Fukunaga K. Introduction to the statistical theory of recognition of images. The translation from English – M.: Glavfizmatizdat, 1979. – 368 pages
52. Feldman Yu.I., Mandurovsky I.A. The theory of fluctuations of the locational signals reflected by the distributed targets. - M.: Radio and Communication, 1988. - 272 pages
53. Gorelik A.G., Chernikov A.A. Some results of a radar research of structure of the wind field at the heights of 50–700 m. Works of the CAO, 1964, issue. 57. - p. 3–18
54. Reference book of climatic characteristics of the free atmosphere on certain stations of the Northern hemisphere. Eds. I.G. Guterman. - M.: NIIAK, 1968
55. Chernikov A.A. Broadening of a spectrum of radar signals from rainfall due to wind shear. Works of the CAO, 1977, issue 126. - p. 48–55
56. Nathanson F.E., Reilly J.P., Cohen M.N. Radar Design Principles. - 2nd Ed. - Mendham, US: SciTech Publishing Inc., 1999. - 720 p
57. Radar systems of air vehicles: Textbook for higher education institutions. Eds. P. S. Davydov. - M.: Transport, 1977. - 352 pages
58. Matveev L.T. Course of the general meteorology. Physics of the atmosphere. - L.: Hydrometeoizdat, 1984. - 751 pages
59. Ryzhkov A.V. Meteorological objects and their radar characteristics. Foreign radio electronics, 1993, No. 4, pages 6–18
60. Ivanov A.A., Melnichuk Yu.V., Morgoyev A.K. The technique of assessment of vertical velocities of air movements in heavy cumulus clouds by means of the doppler radar. Works of the CAO, 1979, issue 135, pages 3–13
61. Vostrenkov V.M., Melnichuk Yu.V. Signals of underlying surface and meteoobjects on the airborne doppler radar. Works of the CAO, 1984, issue 154, pages 52–65
62. Bakulev P.A., Stepin V.M. Methods and devices of selection of moving targets. - M.: Radio and Communication, 1986. - 288 pages
63. Gorelik A.G., Logunov V.F. Determination of velocity of vertical streams in the storm centres and heavy rains at vertical sounding by means of the doppler radar. Works of the CAO, 1972, issue 103, pages 121–133
64. Okhrimenko A.E. Bases of radar-location and radio-electronic fight: Text book for higher education institutions. P.1. Bases of radar-location. - M.: Voyenizdat, 1983. - 456 pages
65. Theoretical bases of a radar-location: Ed. book for higher education institutions. A.A. Korostelev, N.F. Klyuev, Yu.A. Melnik, et al.; Eds. V.E. Dulevich. - M.: Sov. Radio, 1978. - 608 pages
66. Gorelik A.G., Melnichuk Yu.V., Chernikov A.A. Connection of statistical characteristics of a radar signal with dynamic processes and microstructure of a meteoobject. Works of the CAO, 1963, issue 48, pages 3–55
67. Kaplun V.A. Radiotransparent antenna domes. Antennas, issue 8–9 (87–88), 2004, p. 109–116
68. Kravchenko N.I., Lenchuk D.V. Extreme accuracy of measurement of doppler shift of frequency of a meteorological signal when using a pack of coherent signals. News of higher education institutions. Radio electronics, 2001, No. 7, pages 68–80
69. Pyatkin A.K., Nikitin M.V. Realization on FPLD of fast Fourier transformation for DSP algorithms in multipurpose radar stations. Digital processing of signals, 2003, No. 3, pages 21–25
70. Gritsunov A.V. The choice of methods of spectral assessment of temporal functions at modelling of UHF-devices. Radio engineering, 2003, No. 9, pages 25–30
71. Forsyte G., Malkolm M., Mouler To. Machine methods of mathematical calculations. – M.: Mir, 1980
72. The airborne radar for measurement of velocities of vertical movements of lenses in clouds and rainfall. V.M. Vostrenkov, V.V. Ermakov, V.A. Kapitanov et al. Works of the CAO, 1979, issue 135, pages 14–23
73. Veyber Ya.E., Skachkov V.A., Smirnov N.K. Features of measurement of velocities of relative movements of atmospheric formations by radar methods. - M.: SRI OF ACADEMY OF SCIENCES OF THE USSR, 1987. - 13 pages
74. Minkovich B.M., Yakovlev V. P. Theory of synthesis of antennas. - M.: Sov. radio, 1969. - 296 pages

75. Komarov V.M., Andreyeva T. M., Yanovitsky A.K. Airborne pulse-doppler radar stations. Foreign radio electronics, 1991, No. 9–10
76. The reference book on radar-location. Eds. M. Skolnik; The translation from English (in four volumes) Eds. K.N. Trofimov. - V.1. Radar-location bases. Eds. Ya.S. Itskhoka. - M.: Sov. Radio, 1976. - 456 pages
77. Zeger A.E., Burgess L.R. An adaptive AMTI radar antenna array. Proc. IEEE Nat. Aerospace and Electron. Conf. (NAECON'74), Dayton, 1974, New York, 1974, p. 126–133
78. Vorobiev V.G., Zyl V. P., Kuznetsov S.V. Complexes of digital flight navigation equipment. P. 2. Complex of standard flight navigation equipment of the Tu-204 plane. - M.: MGTU GA, 1998 - 116 pages
79. Radar stations of the view of Earth. G.S. Kondratenkov, V.A. Potekhin, A.P. Reutov, Yu.A. Feoktistov; Eds. G.S. Kondratenkov. - M.: Radio and Communication, 1983. - 272 pages
80. Radar stations with digital synthesizing of an antenna aperture. V.N. Antipov, V.T. Goryainov, A.N. Kulin et al.; Eds. V.T. Goryainov. - M.: Radio and Communication, 1988. - 304 pages
81. The guide to forecasting of weather conditions for aircraft. Eds. K.G. Abramovich and A.A. Vasilyev. - L.: Hydrometeoizdat, 1985. - 301 pages
82. Melnik Yu. A., Melnikov V.M., Ryzhkov A.V. Possibilities of use of the single doppler radar in the meteorological purposes: Review Works of GGO, 1991, issue 538. P. 8–18
83. Melnikov V.M. Information processing in doppler MRL. Foreign radio electronics, 1993, No. 4, pages 35–42
84. Banakh V.A., Verner X., Smalikho I.N. Sounding of turbulence of clear sky by a doppler lidar. Numerical modelling. Optics of the atmosphere and ocean, 2001, v. 14, No. 10, pages 932–939

Chapter 4
Conclusion

The monograph is devoted to improvement of methods and algorithms of processing of signals in airborne aviation radar systems (RS) for increase in the flight operation safety (FOS) in difficult meteoconditions.

One of the major factors determining the FOS level is availability of exact and timely information on weather conditions on the route of flight.

The problem of timely and reliable detection of dangerous zones of the meteorological objects (MO) in a front hemisphere at big and average range, measurement of their polar coordinates (range and azimuth) and also the assessment of the degree of their danger is one of the most important for airborne RS of AV of different function. This problem is solved in operation by the digital coherent processing of the reflected signals providing increase in the accuracy of assessment of MO danger for a flight of an AV.

The following main received scientific results are stated in the monograph:

1. The analytical ratios for spectral and power characteristics of the signal reflected by MO and accepted by the airborne RS installed on the mobile carrier in the mode of detection and assessment of danger of WS and zones of turbulence.
2. The MO models, the movements of the RS carrier and the reflected signal allowing to model a background-target situation at the solution of the problem of detection and assessment of danger of zones of WS and turbulence. At the same time the mathematical model of the signal reflected by MO and accepted by an airborne RS, considers the parameters of the probing signal, the physical parameters of the object (water content and radar reflectivity, spatial distribution of wind speed on MO volume, existence of wind shear and turbulence) and also parameters of the movement of the RS carrier (speed, course, existence of trajectory irregularities and elastic fluctuations of the case). The analytical dependences connecting model parameters with R observation conditions are analysed.
3. The algorithms of coherent digital processing of signals in airborne RS directed to improvement of observability and accuracy of assessment of danger of areas of wind shear and turbulence in MO: the parametric modified covariant algorithm of assessment of average frequency and RMS width of a doppler spectrum of

Vereshchagin A. V. et al., *Signal Processing of Airborne Radar Stations*,
Springer Aerospace Technology, https://doi.org/10.1007/978-981-13-9988-6_4

the reflected signal on the basis of the autoregression (AR) model, the algorithm of compensation of own movement of the RS carrier and also the algorithms of assessment of the three-dimensional spatial field of wind speed and definition of degree of danger of the found areas of heavy wind shear and atmospheric turbulence.

The results of researches showed a possibility of increase in accuracy and reliability of detection and assessment of danger of zones of heavy WS and intensive atmospheric turbulence at realization of parametric algorithms of processing of the reflected signals in airborne coherent RS that provides the corresponding increase in the FOS level in general.

The received results can be used during creation of perspective and the modernizations of existing airborne RS of AV of different function.

Annex 1

Characteristics of the main forms of clouds for extra tropic latitudes

Form of clouds	Height of the lower bound, km	Thickness, km	Horizontal dimensions, km	Phase state	Life-time	Average vertical velocities, ω, m/s	Precipitation
Stratoform type clouds							
St	0.1–0.7	0.1–1.0	10–1000	Droplet	A day or more	0.01–0.1	No or small rain
Sc	0.4–2.0	0.1–1.0	10–1000	,,	The same	0.01–0.1	The same
Ac	2.0–6.0	0.1–0.8	10–100	Droplet, mixed	,,	0.01–0.1	No
Cc	6.0–9.0	0.2–1.0	10–100	Crystalline	,,	0.01–0.1	,,
Ns	0.1–2.0	Up to several km	100–1000	Mixed	,,	0.01–0.1	Rain, snow
As	3.0–6.0	The same	100–1000	Mixed, crystalline	,,	0.01–0.1	The same
Cs	5.0–9.0	,,	100–1000	Crystalline	,,	0.01–0.1	No
Ci	6.0–10.0	From tens of meters up to 1.0–1.5 km	10–1000	,,	,,	0.01–0.1	,,
Cumuloform clouds							
Cu	0.8–2.0		1.0–5.0	Droplet	Tens of minutes	1	No
Cu cong.	0.8–2.0		5.0–10.0	,,	,,	1–50	,,
Cb	0.4–1.5		Up to 50–100	Mixed	Tens of minutes, sometimes hours	5–10	Rain, storm

Notes 1. The values ω for Cu, Cb belong to average velocities of the intra cloudy ascending flows. The velocity of the descending flows sometimes is by 1.5–2 times less, than of ascending ones.
2. The data on thickness of Cu and Cb belong to summer clouds of midlatitudes. The cumuliform clouds which are observed at whiter fronts have considerably smaller thickness.

Vereshchagin A. V. et al., *Signal Processing of Airborne Radar Stations*,
Springer Aerospace Technology, https://doi.org/10.1007/978-981-13-9988-6

Annex 2

The Model of a "Ring Vortex" of the Wind Field in the Descending Air Flow

Let us consider the case of the single isolated descending flow which can be presented in the form of three-dimensional vortex field with a toroidal core (Fig. A2.1), symmetric in relation to the Z' axis.

The vortex radius—R_V, the core radius—R_0, the height over underlying surface—z_V, the velocity in a flow central part—V_V. Let the point P be any point out of care located at r_{PV} distance from the axis Z' at the height z_P.

Then the function of the flow of the vector field Φ of the single isolated circular vortex through a circle (z_P, r_P)

$$\psi = -\Psi/2\pi = -r_{PV}F,$$

is connected with the values of projections of wind speed in the point P by the ratios

$$v_z = -\frac{1}{r_{PV}}\frac{\partial\psi}{\partial r_{PV}}, \quad v_r = \frac{1}{r_{PV}}\frac{\partial\psi}{\partial z}, \tag{A2.1}$$

being the sequences of the equation of the flow continuity. In the expression (A2.1)$\Psi = 2\pi r_{PV}F$—the flow of the vector field; v_z and v_r—the wind speed projections in the directions Z' and r_{PV}.

The value of the flow function in the point (z_P, r_{PV}), which is caused by one vortex thread with the circulation ν and the coordinates (Z', R_V) (we will notice that the element of vortex thread has the length $R_V d\theta$, where θ—the element angle with the direction F) is equal to

$$\psi = -r_{PV}\Phi = -\frac{\nu r_{PV}R_V}{4\pi}\int_0^{2\pi}\frac{\cos\theta}{r}d\theta,$$

Vereshchagin A. V. et al., *Signal Processing of Airborne Radar Stations*, Springer Aerospace Technology, https://doi.org/10.1007/978-981-13-9988-6

where $r = \sqrt{(z - z_P)^2 + r_{PV}^2 + R_V^2 - 2r_{PV}R_V\cos\theta}$.

Let's designate through r_1 and r_2 respectively the smallest and the biggest distances from the point P to vortex points

$$r_1^2 = (z_P - z_V)^2 + (r_{PV} - R_V)^2, \quad r_2^2 = (z_P - z_V)^2 + (r_{PV} + R_V)^2.$$

Then $r^2 = r_1^2\cos^2\frac{\theta}{2} + r_2^2\sin^2\frac{\theta}{2}$, $4r_{PV}R_V\cos\theta = r_1^2 + r_2^2 - 2r^2$, and, therefore

$$\psi = -\frac{\nu}{8\pi}\left[\left(r_1^2 + r_2^2\right)\int\limits_0^{\pi}\frac{d\theta}{\sqrt{r_1^2\cos^2\frac{\theta}{2} + r_2^2\sin^2\frac{\theta}{2}}} - 2\int\limits_0^{\pi}\sqrt{r_1^2\cos^2\frac{\theta}{2} + r_2^2\sin^2\frac{\theta}{2}}d\theta\right]. \tag{A2.2}$$

the integrals in (A2.2) belong to the type of full elliptic integrals which values can be approximately determined by numerical methods. If we set

$$\begin{cases} k = \frac{r_2 - r_1}{r_2 + r_1}, \\ F_1(k) = \frac{(r_1^2 + r_2^2)}{8\pi(r_1 + r_2)}\int\limits_0^{\pi}\frac{d\theta}{\sqrt{r_1^2\cos^2\frac{\theta}{2} + r_2^2\sin^2\frac{\theta}{2}}}, \\ E_1(k) = \frac{1}{4\pi(r_1 + r_2)}\int\limits_0^{\pi}\sqrt{r_1^2\cos^2\frac{\theta}{2} + r_2^2\sin^2\frac{\theta}{2}}d\theta. \end{cases}$$

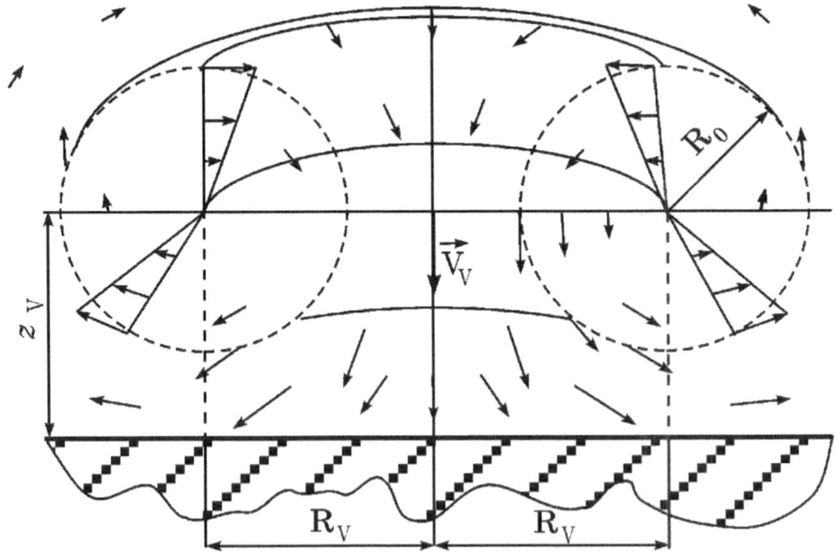

Fig. A2.1 The spatial distribution of the wind speed in the descending air flow

then, using the Landen transformation, the expression (A2.2) for the flow ψ can be brought to the form

$$\psi = -\frac{\nu}{2\pi}(r_1 + r_2)[F_1(k) - E_1(k)]. \tag{A2.3}$$

For approximation of a combination of elliptic integrals (A2.3) it is possible to use the expression of the form

$$[F_1(k) - E_1(k)] \approx \frac{0.788k^2}{0.25 + 0.75\sqrt{1-k^2}}. \tag{A2.4}$$

As the vortex field is located in close proximity to the underlying surface (US), the latter significantly influences the movement of air around a vortex. In [1] it is shown that in this case the US influence can be presented in the form of the second vortex field located in relation to the US in an inversed manner to the first real vortex (Fig. A2.2). Then the flow function (A2.3) can be written down in the form of superposition of the corresponding functions for two vortexes operating at the same time

$$\psi = -\frac{\nu}{2\pi}\{(r_{11} + r_{12})[F_{11}(k_1) - E_{11}(k_1)] - (r_{21} + r_{22})[F_{12}(k_2) - E_{12}(k_2)]\}, \tag{A2.5}$$

where

$r_{11} = \sqrt{(z_P - z_V)^2 + (r_{PV} - R_V)^2}$—the smallest distance from the point P to points of a real vortex;

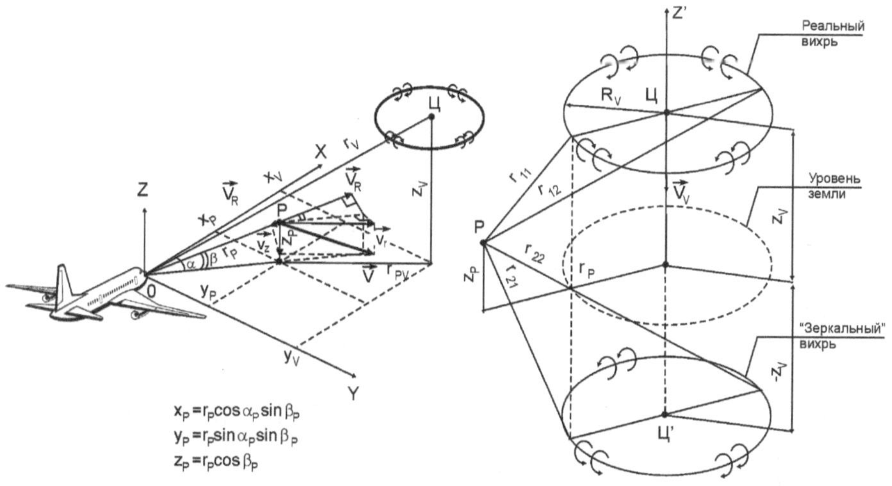

Fig. A2.2 Proceeding from the geometrical ratios

$r_{12} = \sqrt{(z_P + z_V)^2 + (r_{PV} - R_V)^2}$—the biggest distance from the point P to points of a real vortex;
$r_{21} = \sqrt{(z_P - z_V)^2 + (r_{PV} + R_V)^2}$—the smallest distance from the point P to points of a mirror vortex;
$r_{22} = \sqrt{(z_P + z_V)^2 + (r_{PV} + R_V)^2}$—the biggest distance from the point P to points of a mirror vortex;
$F_{11}(k_1), E_{11}(k_1)$—the values of elliptical integrals for a real vortex;
$F_{12}(k_2), E_{12}(k_2)$—the values of elliptical integrals for a mirror vortex;

$$k_1 = \frac{r_{12} - r_{11}}{r_{12} + r_{11}}; \quad k_2 = \frac{r_{22} - r_{21}}{r_{22} + r_{21}}.$$

Taking into account the approximation (A2.4) the expression (A2.5) is brought to the form

$$\psi = -\frac{v}{2\pi}\left\{(r_{11} + r_{12})\frac{0.788k_1^2}{0.25 + 0.75\sqrt{1 - k_1^2}} - (r_{21} + r_{22})\frac{0.788k_2^2}{0.25 + 0.75\sqrt{1 - k_2^2}}\right\}. \tag{A2.6}$$

The circulation v of the vortex field is connected with the velocity V_V in its central part by the ratio [1]

$$v = \frac{2V_V R_V}{1 - \frac{1}{\left[1 + \left(\frac{2z_V}{R_V}\right)^2\right]^{1.5}}}. \tag{A2.7}$$

The vertical and radial* components of the wind speed in the point P which is out of the vortex core according to (A2.1) can be received by numerical differentiation of the flow function

$$v_z = -\frac{1}{r_{PV}}\frac{\partial\psi}{\partial r_{PV}} = \frac{\psi(z_P, r_{PV}) - \psi(z_P, r_{PV} + \Delta r)}{r_{PV}\Delta r}, \tag{A2.8}$$

$$v_r = \frac{1}{r_{PV}}\frac{\partial\psi}{\partial z} = \frac{\psi(z_P + \Delta z, r_{PV}) - \psi(z_P, r_{PV})}{r_{PV}\Delta z}. \tag{A2.9}$$

*Note: *The radial component v_r of velocity in this case represents a wind speed projection in the point P on the direction r_P.*

Projections of the radial component of the wind speed in the point P on the *X* and *Y* axis of the BCS (the longitudinal and cross components of wind speed respectively) can be defined by the ratios (Fig. A2.2)

$$v_x = v_r \frac{(x_V - x_P)}{r_{PV}}, \tag{A2.10}$$

$$v_y = v_r \frac{(y_V - y_P)}{r_{PV}}, \tag{A2.11}$$

where

(x_V, y_V, z_V) —the coordinates of the vortex centre in the BCS;
(x_P, y_P, z_P) —the coordinates of the point P in the BCS

Passing in the BCS from the Cartesian coordinates to polar coordinates, we will write down the coordinates of the centre of a circular vortex in the form

$$\begin{cases} x_V = r_V \cos\alpha_V \sin\beta_V, \\ y_V = r_V \sin\alpha_V \sin\beta_V, \\ z_V = r_V \cos\beta_V, \end{cases} \tag{A2.12}$$

and also the coordinates of the point *P* (the centre of the allowed volume) in the form

$$\begin{cases} x_P = r_P \cos\alpha_P \sin\beta_P, \\ y_P = r_P \sin\alpha_P \sin\beta_P, \\ z_P = r_P \cos\beta_P. \end{cases} \tag{A2.13}$$

where

(r_V, α_V, β_V) —the range, azimuth and elevation angle of the vortex centre in the BCS;
(r_P, α_P, β_P) —the range, azimuth and elevation angle of the centre of the allowed volume in the BCS

Within the vortex core the wind speed equal to zero in the centre, linearly increases on radius to the core border (Fig. A2.1). Therefore for the points within the core (i.e. at $\sqrt{(x_P - x_V)^2 + (y_P - y_V)^2 + (z_P - z_V)^2} \le R_0$) it can be found by recalculation from the core in proportion to distance to the core centre

$$v = v_{\max} \frac{\sqrt{(x_P - x_V)^2 + (y_P - y_V)^2 + (z_P - z_V)^2}}{R_0}, \tag{A2.14}$$

where

$$v_{\max} = \frac{v}{2R_V}\left\{\frac{1}{\left[1+\left(\frac{z_V-z_P}{R_V}\right)^2\right]^{1,5}} - \frac{1}{\left[1+\left(\frac{-z_V-z_P}{R_V}\right)^2\right]^{1,5}}\right\} \tag{A2.15}$$

—the maximum velocity on border of the core [1].

Radial (in the "point–vortex centre" direction) and vertical projections of the wind speed (A2.14) in the point P which is within the core are equal respectively

$$v_r = v\frac{z_P - z_V}{\sqrt{(x_P - x_V)^2 + (y_P - y_V)^2 + (z_P - z_V)^2}} = v_{\max}\frac{z_P - z_V}{R_0}, \tag{A2.16}$$

$$v_z = v\frac{r_P - r_V}{\sqrt{(x_P - x_V)^2 + (y_P - y_V)^2 + (z_P - z_V)^2}} = v_{\max}\frac{r_P - r_V}{R_0}. \tag{A2.17}$$

The expression (A2.16) and (A2.17) describe three-dimensional spatial fields of radial and vertical components of wind speed in the descending flow in the form of a "ring vortex" (the ring-vortex model).

The airborne RS in the course of observation determines the radial wind speed V_R in the point P as a wind speed projection to the direction connecting APC to the specified point. Proceeding from the geometrical ratios given the in Fig. A2.2, the value V_R is equal to the sum of projections of velocities v_r (A2.16) and v_z (A2.17) to the direction connecting the points O (ARC) and P (the centre of the allowed volume).

At first we determine the projection of the radial component v_r (A2.16) of wind speed on the vertical plane passing through the points O and P,

$$v_{rv} = v_r \cos|\alpha - \alpha'| = v_r \cos\left|\alpha - \mathrm{arctg}\frac{y_0 - y_P}{x_0 - x_P}\alpha\right|,$$

where α'—the angle between the axis X and the r_P direction.

Further we design the velocities v_z and v_{rv} on the direction connecting the points O and P. The sum of the received projections will give as a result the required value of radial velocity of the centre of the allowed volume in relation to APC

$$V_R = v_r \cos\left|\alpha_P - \mathrm{arctg}\frac{y_V - y_P}{x_V - x_P}\alpha_P\right| \cos\beta_P + v_z \sin\beta_P. \tag{A2.18}$$

Annex 3

Parametric Models of the Reflected Signal

In many practical cases the target situation causing formation of the reflected signal can be rather precisely simulated by the linear system of a finite order. Generally the connection between input and output signals in a linear system (the forming filter) with the rational transfer function is described by the linear differential equation

$$s(m) = \sum_{k=1}^{p} a_k s(m-k) + \sum_{k=1}^{q} b_k e(m-k) + e(m), \qquad \text{(A3.1)}$$

where $s(m), e(m)$—the output and entrance system signals respectively.

This model is known as the model of autoregression—the moving average (ARMA) [2] (Fig. A3.1a). Hereinafter for improvement of presentation and simplification of record of expressions we will omit the multiplier T_p at the indication of number of reading of a random process. In the expression (A3.1): p—the order of autoregression (AR); q—the order of moving average (MA); a_k—complex AR coefficients; b_k—complex MA coefficients.

If all coefficients $b_k = 0$, then the signal *s(m)* represents a linear regression of its previous values

$$s(m) = \sum_{k=1}^{p} a_k s(m-k) + e(m). \qquad \text{(A3.2)}$$

Such a model is called the autoregression model (Fig. A3.1b). Respectively, if all coefficients $a_k = 0$, then this is the *model of moving average* (the Fig. A3.1c).

The Eq. (A3.2) can be written down in a matrix form

$$\vec{a}^T \vec{S}(m) = \vec{b}^T \vec{E}(m), \qquad \text{(A3.3)}$$

Vereshchagin A. V. et al., *Signal Processing of Airborne Radar Stations*,
Springer Aerospace Technology, https://doi.org/10.1007/978-981-13-9988-6

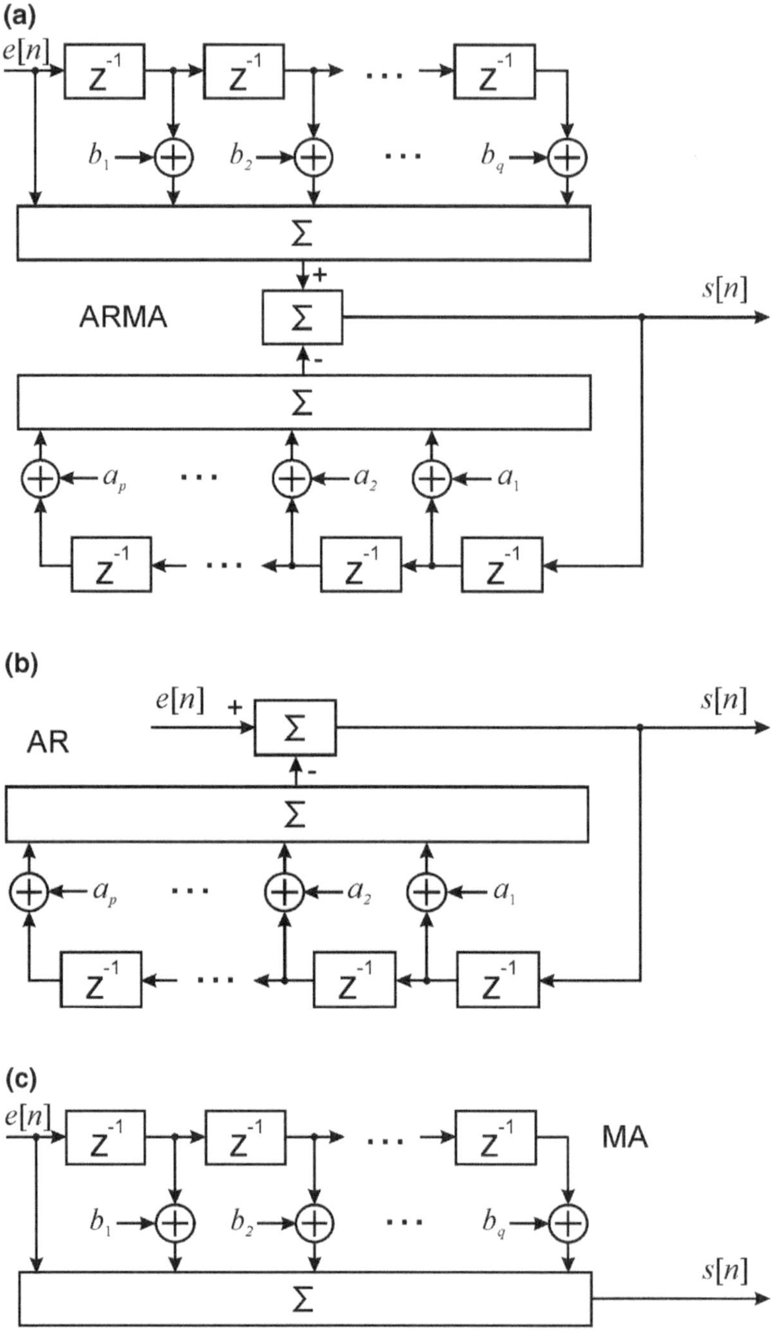

Fig. A3.1 Block diagrams of forming filters with rational transfer function: **a** for the ARMA model; **b** for the AR model; **c** for the MA model

where

T –the transposing operation;

$\vec{a} = \begin{bmatrix} 1 \\ -a_1 \\ \cdots \\ -a_p \end{bmatrix}$ –$(p + 1)$—the element vector of AR coefficients;

$\vec{S}(m) = \begin{bmatrix} s(m) \\ s(m-1) \\ \cdots \\ s(m-p) \end{bmatrix}$ –(p + 1)—the element vector of readings of an output signal;

$\vec{b} = \begin{bmatrix} 1 \\ b_1 \\ \cdots \\ b_q \end{bmatrix}$ –$(q + 1)$—the element vector of MA coefficients;

$\vec{E}(m) = \begin{bmatrix} e(m) \\ e(m-1) \\ \cdots \\ e(m-q) \end{bmatrix}$ –$(q + 1)$—the element vector of readings of an input signal

The Transfer function (TF) of the considered linear system can be presented by means of the technique of z-transformations [3, 4]. Applying z-transformation to both members of the Eq. (A3.1), considering properties of linearity and shift [3], we receive

$$S(z) = \sum_{k=1}^{p} a_k z^{-k} S(z) + \sum_{k=1}^{q} b_k z^{-k} E(z) + E(z), \tag{A3.4}$$

where $z = \exp(j2\pi f \tau)$;

$S(z) = Z\{s(m)\} = \sum\limits_{k=0}^{\infty} s(k) z^{-k}$ —the z-image of an output system signal;

$E(z) = Z\{e(m)\} = \sum\limits_{k=0}^{\infty} e(k) z^{-k}$ —the z-image of the input (forming) system signal.

It follows from the expression (A3.4)

$$S(z) = H(z)E(z),$$

where $H(z) = b(z)/a(z)$—the TF of the considered linear system;

$$b(z) = 1 + \sum_{k=1}^{q} b_k z^{-k}; \quad a(z) = 1 - \sum_{k=1}^{p} a_k z^{-k}.$$

Let the input signal *e(m)* of the system be δ-function $e(m) = \delta(m)$, then the output system signal will represent its pulse characteristic. As at the same time $E(z) = 1$, then $S(z) = Z\{h(m)\}$, therefore,

$$H(z) = Z\{h(m)\} = 1 + \sum_{k=1}^{\infty} h(k)z^{-k}, \tag{A3.5}$$

that is the TF is the z-image of system impulse characteristic.

If we multiply both members of the Eq. (A3.1) by $s^*(m-n)$ and average the received result, then we receive

$$\begin{aligned}\overline{s(m)s^*(m-n)} &= \sum_{k=1}^{p} a_k \overline{s(m-k)s^*(m-n)} + \\ &+ \sum_{k=1}^{q} b_k \overline{e(m-k)s^*(m-n)} + \overline{e(m)s^*(m-n)},\end{aligned} \tag{A3.6}$$

or

$$B(n) = \sum_{k=1}^{p} a_k B(n-k) + \sum_{k=1}^{q} b_k B_{es}(n-k) + B_{es}(n). \tag{P3.7}$$

As the output signal of the system is equal to convolution of the input signal with pulse characteristic, the mutual correlation function $B_{es}(n)$ between output and input signals of the system can be written down with use of readings of pulse characteristic (A3.5) [5]

$$\begin{aligned}B_{es}(n) &= \overline{e(m+n)s^*(m)} = \overline{e(m+n)\left[e^*(m) + \sum_{k=1}^{p} h^*(k)s^*(m-k)\right]} \\ &= B_e(n) + \sum_{k=1}^{p} h^*(k)B_e(n+k),\end{aligned}$$

where $B_e(n)$—AKC of the input signal *e(m)*.

In case of use as an excitation signal *e(m)* of discrete complex normal white noise with zero average value and dispersion σ_e^2

$$B_{es}(n) = \begin{cases} 0, & n > 0; \\ \sigma_e^2, & n = 0; \\ \sigma_e^2 h^*(-n), & n < 0. \end{cases}$$

From here the expression connecting the ARMA model parameters with ACF of the output signal s(m) of the linear system [5, 6] follows

$$B(m)=\begin{cases} B^*(-m), & m<0; \\ \sum\limits_{k=1}^{p} a_k B(m-k)+\sigma_e^2 \sum\limits_{k=m}^{q} b_k h^*(k-m), & 0\le m\le q; \\ \sum\limits_{k=1}^{p} a_k B(m-k), & m>q. \end{cases} \quad \text{(A3.8)}$$

Thus, when using the linear system with the rational transfer function *H(z)* the setting of p of the ACF consecutive values of the *s(m)* signal allows to continue it unambiguously indefinitely by means of the recurrence ratio

$$B(m)=\sum_{k=1}^{p} a_k B(m-k) \text{ For all } m>q, \quad \text{(A3.9)}$$

Setting q = 0, the Eq. (A3.9) is transformed to the ratio connecting the AR model parameters of the output signal of the system with its values of ACF

$$B(m)=\begin{cases} B^*(-m), & m<0; \\ \sum\limits_{k=1}^{p} a_k B^*(k)+\sigma_e^2, & m=0; \\ \sum\limits_{k=1}^{p} a_k B(m-k), & m>0. \end{cases} \quad \text{(A3.10)}$$

Let p = 0, then taking into account that $h[k]=b[k]$ at $1\le k\le q$, the equation (A3.8) is transformed to the ratio connecting the MA model parameters of the reflected signal with its values of ACF

$$B(m)=\begin{cases} B^*(-m), & m<0; \\ \sigma_e^2 \sum\limits_{k=m}^{q} b_k b_{k-m}^*, & 0\le m\le q; \\ 0, & m>q. \end{cases} \quad \text{(A3.11)}$$

It follows from the expression (A3.11) that ACF and the MA coefficients are connected by nonlinear dependence of convolution type.

It should be noted that at Gaussian approach infinite continuation of ACF on its values, assumed by the ARMA model, maximizes the entropy

$$h_{y\partial}=\frac{1}{4f_{max}}\int\limits_{-f_{max}}^{f_{max}} \ln[S(f)]df, \quad \text{(A3.12)}$$

of corresponding time sequences (sequences of readings of signals) [7, page 53].

Here f_{max}—the greatest frequency component in a signal spectrum.

In other words, the use of the ARMA model is equivalent to application of the method of the entropy maximum (EM) [5, 6, 8].

Let the first 2 M + 1 ACF values be known. Then the most rational way of determination of its unknown values is that new values do not increase the available information or do not reduce entropy of the considered signal [7], i.e.

$$\frac{\partial h_d}{\partial B(m)} = 0, \quad when\ |m| \geq M+1, \tag{A3.13}$$

Taking into account (A3.9), (A3.12) and (A3.13) in case of use as an excitation signal *e(m)* of white noise with zero average value and the dispersion σ_e^2 the assessment of spectral density of power of the *s(m)* signal is transformed to the form [9, 10]

$$S(f) = \sigma_e^2 T_n \frac{\left|1 + \sum_{k=1}^{q} b_k \exp(-j2\pi fkT_n)\right|^2}{\left|1 - \sum_{k=1}^{p} a_k \exp(-j2\pi fkT_n)\right|^2}, \tag{A3.14}$$

It follows from the expression (A3.14) that for the assessment of *S(f)* it is necessary to define the values a_k, b_k of the AR and MA coefficients and also the dispersion σ_e^2 of an excitation signal (noise). The assessment of the specified parameters can be made by approximation of the set power spectrum of a signal or by processing selection of real signals [11].

The expression (A3.14) can be presented in the vector form

$$S(\omega) = \sigma_e^2 T_n \frac{\vec{c}_q^H(f)\vec{b}\vec{b}^H\vec{c}_q(f)}{\vec{c}_p^H(f)\vec{a}\vec{a}^H\vec{c}_p(f)} \tag{A3.15}$$

where $\vec{c}_q(f) = \begin{bmatrix} 1 \\ \exp(j2\pi fT_n) \\ \cdots \\ \exp(j2\pi fqT_n) \end{bmatrix}$, $\vec{c}_p(f) = \begin{bmatrix} 1 \\ \exp(j2\pi fT_n) \\ \cdots \\ \exp(j2\pi fpT_n) \end{bmatrix}$—the vectors of complex sinusoids;

N—the symbol of the operation of Hermite conjugation consisting in consecutive performance of operations of transposing and complex conjugation.

Setting $q = 0$, the expression (A3.14) is transformed to the ratio for spectral density of the AR model

$$S(f) = \sigma_e^2 T_n \left|1 - \sum_{k=1}^{p} a_k \exp(-j2\pi fkT_n)\right|^{-2} \tag{A3.16}$$

or in a vector form

$$S(f) = \frac{\sigma_e^2 T_n}{\vec{c}_p^H(f)\vec{a}\vec{a}^H\vec{c}_p(f)}. \tag{A3.17}$$

Let's factorize the polynom $A(f) = 1 + \sum_{k=1}^{p} a_k \exp(-j2\pi fkT_n)$, which is in the denominator of the spectral density (A3.16). For this purpose we execute the z-transformation of the form $z = \exp(j2\pi fT_n)$ and find the polynom roots $A(z)$, having solved the corresponding characteristic equation

$$A(z) = 1 + \sum_{k=1}^{p} a_k z^{-k} = \sum_{k=0}^{p} a_k z^{-k} = z^{-p} \sum_{k=1}^{p} a_k z^{p-k} = 0, \tag{A3.18}$$

where $a_0 = 1$.

According to the main theorem of algebra the polynom of the p-th power with complex coefficients has p complex roots $z_k = |z_k| \exp(j\phi_k)$ (taking into account their frequency rate) called the poles of the AR model. From this it follows that (A3.18) can be presented in the form

$$A(z) = z^{-p} \prod_{k=1}^{p} (z - z_k) = \prod_{k=1}^{p} \left(1 - \frac{z_k}{z}\right).$$

As a result of the inverse z-transformation

$$A(f) = \prod_{k=1}^{p} \{1 - |z_k| \exp[-j(2\pi fT_n - \phi_k)]\} = \prod_{k=1}^{p} A_k(f),$$

where $A_k(f) = 1 - |z_k| \exp[-j(2\pi fT_n - \phi_k)]$.

If we present $A_k(f)$ in the exponential form

$$A_k(f) = |A_k(f)| \exp[-j\varphi_{Ak}],$$

where $\varphi_{Ak} = \mathrm{Arg}[A_k(f)] = \mathrm{arctg}\frac{\mathrm{Im}[A_k(f)]}{\mathrm{Re}[A_k(f)]}$, the spectral plane (A3.16) can be written down in the form

$$S(f) = \frac{\sigma_e^2 T_n}{\left| \left(\prod_{k=1}^{p} |A_k(f)| \right) \exp\left[j \sum_{k=1}^{p} \varphi_{Ak} \right] \right|^2} = \frac{\sigma_e^2 T_n}{\left(\prod_{k=1}^{p} |A_k(f)| \right)^2}. \tag{A3.19}$$

Definition of poles of the AR model, especially of low (second or third) order [7, 12, 13], can be made by one of the methods of calculation of roots of polynoms (the method of the analysis of own values of the attached matrix, Laguerre, Lobachevsky methods, etc.).

If we use the AR process of the first order as a model of the MO signal, then its spectrum (A3.16) symmetric in relation to $\bar{f}$, can be presented in the form

$$S(f) = \frac{\sigma_e^2 T_n}{|1 - a_1 \exp(-j2\pi f T_n)|^2} = \frac{\sigma_e^2 T_n}{1 + |a_1|^2 - 2|a_1|\cos(2\pi f T_n - F)},$$

where $a_1 = |a_1|\exp(jF)$—the unknown complex coefficient of the AR model.

If the AR coefficient module $|a_1| \to 1$, then the modelled signal is the narrow-band random sequence which spectrum average frequency coincides with the spectrum maximum frequency [14]

$$\bar{f} = F/(2\pi T_n) = Arg(a_1)/(2\pi T_n), \qquad \text{(A3.20)}$$

and the expression for the spectrum RMS width after a number of transformations is brought to the form [15]

$$\Delta f = \frac{1}{2\pi T_n}\left[\frac{\pi^2}{3} - 4\sum_{k=1}^{\infty}\frac{(-1)^{k+1}}{k^2}|a_1|^k\right]^{1/2}. \qquad \text{(A3.21)}$$

Thus, the assessment of AR coefficient a_1 is enough for measurement of $\bar{\omega}$ and $\Delta\omega$.

At $p = 0$ from (A3.14) we receive the expression for the spectral density of the MA model in the vector form

$$S(f) = \sigma_e^2 T_n \vec{c}_q^H(f)\vec{b}\vec{b}^H\vec{c}_q(f).$$

The typical spectra of the ARMA, AR and MA processes are given in Fig. A3.2 [5, page 220]. The sharp peaks are characteristic of spectra of the AR processes, and deep holes—of the spectra of the MA processes. Thus, the MA models are unsuitable for modelling of spectra of narrow-band signals. It is expedient to use them for spectral estimation of the processes which spectra have wide maxima or narrow holes (minimal, zero). The spectrum of the ARMA process representing the result of association of the spectra of the AR and MA processes is suitable for modelling of spectra of the real signals having both sharp peaks, and deep holes.

One of the most important parameters of the ARMA, AR and MA is their order. Provision of a compromise between resolution and accuracy (dispersion) of the received spectral assessments depends on the choice of the order [5, page 255]. If the order of model is chosen as too small, then strongly smoothed spectral estimates are received. For decrease in dispersion of estimates in this case it is necessary to accumulate big selections of signals (about 1000) that is unacceptable in case of use of an airborne RS. The overdetermination of the model (excessive overestimation of an order) in the conditions of existence of errors of measurement can lead to additional, often very essential, errors in the assessment of the spectrum [5, 16]. In particular, emergence of false maxima in the spectrum is possible.

Several various criteria (objective functions) are offered for the choice of the model order [5, 17].

1. *Akaike Information Criterion* (AIC)

According to this criterion based on the technique of maximum likelihood, the choice of the model order is carried out by minimization of some theoretical-information function

$$p^t = \min_p AIC(p),$$

$$AIC(p) = -2\max L_p + 2I(p) \tag{A3.22}$$

where $\max L_p$—the maximum of a logarithm of likelihood function at the fixed value of the order p;

$I(p)$—the number of the independent parameters defining the model (P3.2). In the assumption of gaussianity of statistics of the studied process the function (A3.22) can be defined by the following expression [5]

$$AIC(p) = M \ln \hat{\hat{\sigma}}_e^2 + 2p, \tag{A3.23}$$

where M—the length of the processed pack of readings;

$\hat{\hat{\sigma}}_e^2$—the assessment of value of dispersion of the forming noise in (P3.1). The disadvantage of this criterion is the absence of aspiration to zero of an error of assessment at choosing of the correct order p and at $M \to \infty$, and it leads to significant overestimation of the model order in case of long packs [18].

2. *Minimal Description Length Criterion* (MDL)

For elimination of noted disadvantage *J. Rissanen* modified the AIC criterion, having presented it in the form [18]

$$\hat{p} = \min_p MDL(p),$$

where

$$MDL(p) = -2\max L_p + I(p)\ln M.$$

For the reflected analysed Gaussian signal the function $MDL(p)$ can be described by the expression

$$MDL(p) = M \ln \hat{\hat{\sigma}}_e^2 + p \ln M. \qquad (A3.24)$$

3. *Maximum Posteriori Probability Criterion* (MPP)

The maximum posteriori probability criterion consists in minimization of function of the type [17, p. 11]

$$MAP(p) = M \ln \hat{\hat{\sigma}}_e^2 + \frac{5}{3} p \ln M. \qquad (A3.25)$$

4. *Autoregression Transfer Function* (ATF) *Criterion*

In this case the order of the model is chosen as equal to the order at which the assessment of difference of an average square of errors between the model and the modelled process would be minimal

$$\hat{p} = \min_p \left(\frac{1}{N} \sum_{i=1}^{p} \bar{\sigma}_{ei}^{-1} - \bar{\sigma}_{ep}^{-1} \right), \qquad (A3.26)$$

where $\bar{\sigma}_{ei}^{-1} = \frac{N}{N-i} \sigma_{ei}$, σ_{ei}—the RMS error of the assessment on the *i*-th step.

The results of estimation of ranges when using the criteria given above differ insignificantly from each other, especially in case of real data, but not modelled processes with the set of statistical properties [5].

Annex 4

Short Description of the Modelling Software

In the course of the researches conducted within this thesis work the package of application programs providing modelling of the developed algorithms and also calculation of their main characteristics is created. The basis of the package are the mathematical models considered in the Chap. 2 of the thesis work: the MO model in the conditions of WS and turbulence, the model of the movement of the RS carrier, the model of the reflected signal and the model of a processing path. The vector organization of calculations in the *MATLAB* system allowed to use widely at modelling the methods of linear algebra that considerably reduced the time of calculations. The developed package includes the following programs:

Model1—the program of calculation of the main characteristics of the developed algorithm of detection and assessment of the degree of danger of MO zones;
CoeffMQ—the program of calculation of dependence of digits of coefficients of the processing device on volume of the processed pack at the different values of SNR;
ProbDF—the program of calculation of probability of the correct detection of dangerous zones depending on the rated RMS value of fluctuation noise of measurement of parameters of a Doppler spectrum;
ROh—the program of calculation of the dependence of RR assessment error on the height over the MO lower boundary.

All developed modelling programs possess the same interface in the form of the system of graphic context menus. The general modelling scheme provides the following actions:

- the input, viewing and editing of modelling parameters;
- the modelling of MO, the movement of the RS carrier, the reflected signal and a RS processing path;

Vereshchagin A. V. et al., *Signal Processing of Airborne Radar Stations*,
Springer Aerospace Technology, https://doi.org/10.1007/978-981-13-9988-6

- the calculation of the required characteristics (probabilities of the correct detection of potentially dangerous MO, accuracy of assessment of the DS parameters, etc.);
- the analysis of the results (display of numerical values and schedules);
- the record of results in the *mat*-file (the standard data file type in the *MATLAB* system).

The following functions are developed at writing of the programs:

Aquatic—the function of calculation of MO water content (depending on height);
AR_fDf—the function of the assessment of $\bar{f}$ and Δf on the AR-coefficient;
AR_Order—the assessment of the order of the AR-model;
AR_Pole—the function of calculation of a vector of poles of the AR-model;
AR_PSelect—the function of the choice of a pole of the AR-model;
Dnoise—the function of calculation of dispersion of own noise of RS receiver;
Df_Pogr—the function of calculation of errors of the assessment Δf;
LyLz—the function of calculation of shift of an APC in horizontal and vertical directions
Model2—the function of the analysis of dependence of accuracy of compensation of the AV movement on various factors (accuracy of data of accelerometers, TI/EMS, etc.);
ModTNEMS—the function of modelling of TI and EMS of an AV-carrier
ModW—the function of calculation of projections of an AV air velocity
NAE—the function of assessment of number of dephased AE
ParamLA—the function of display of the parameters of an AV-carrier;
ParamLAEdit—the function of change of the parameters of an AV-carrier;
ParamMO—the function of display of the MO parameters;
ParamMOEdit—the function of change of the MO parameters;
ParamRLS—the function of display of the RS parameters;
ParamRLSEdit—the function of change of the RS parameters;
Reflectivity—the function of calculation of dependence of radar reflectivity of the allowed MO volume on height;
SEpr—the function of calculation of total EER of the allowed MO volume;
SEpr1—the function of calculation of dependence of total EER of the allowed MO volume on range;
SigDfCOV—the function of calculation of accuracy of assessment of width of a MO Doppler spectrum by the AR method (the modified covariant algorithm);
SigDfdV—the function of the analysis of dependence of accuracy of assessment of width of a MO Doppler spectrum on width of a spectrum of velocity of atmospheric turbulence;
SigDfFFT—the function of calculation of accuracy of assessment of width of a MO Doppler spectrum by the periodogram method;

SigDfM—the function of the analysis of dependence of accuracy of assessment of width of a MO Doppler spectrum on number of impulses in the processed pack;
SigDfMP—the function of calculation of accuracy of assessment of width of a MO Doppler spectrum by the maximum likelihood method;
SigDfMUSIC—the function of calculation of accuracy of assessment of average frequency of a MO Doppler spectrum by the MUSIC method;
SigDfPI—the function of calculation of accuracy of assessment of width of a MO Doppler spectrum by the method of paired pulses;
SigDVAlpha—the Program of calculation of dependence of an error of assessment of a RMS width of a spectrum of velocities of turbulent MO on AS DP axis deviation angle;
SigDVdV—the function of the analysis of dependence of accuracy of assessment of width of a spectrum of velocities of turbulent MO on its true value;
SigDVM—the function of the analysis of dependence of accuracy of assessment of width of a spectrum of velocities of a turbulent MO on number of impulses in the processed pack;
SigfAxe—the function of calculation of an error of assessment of an AV velocity due to drift of the accelerometer;
SigfdV—the function of the analysis of dependence of accuracy of assessment of an average frequency of a MO Doppler spectrum on width of a spectrum of velocity of atmospheric turbulence;
SigfCOV—the function of calculation of accuracy of assessment of an average frequency of a MO Doppler spectrum by the AR method (the modified covariant algorithm);
SigfFFT—the function of calculation of accuracy of assessment of an average frequency of a MO Doppler spectrum by the periodogram method;
SigfM—the function of the analysis of dependence of accuracy of assessment of an average frequency of a MO Doppler spectrum on number of impulses in the processed pack;
SigfMP—the function of calculation of accuracy of assessment of an average frequency of a MO Doppler spectrum by the maximum likelihood method;
SigfMUSIC—the function of calculation of accuracy of assessment of an average frequency of a MO Doppler spectrum by the MUSIC method;
SigfPT—the function of calculation of accuracy of assessment of an average frequency of a MO Doppler spectrum by the method of paired pulses;
SigVdV—the function of the analysis of dependence of accuracy of assessment of an average velocity of a MO allowed volume on width of a spectrum of velocity of atmospheric turbulence;
SigVM—the function of the analysis of dependence of accuracy of assessment of an average velocity of a MO allowed volume on number of impulses in the processed pack;
Stream—the function of calculation of value of a ring vortex flow;

Turbulence—the function of calculation of velocity of dissipation of kinetic energy of turbulence;
VFK – the function of calculation of a vector of phase correction for compensation of radial shift of an AV in relation to a MO;
Wind1—the function of calculation of the direction of wind on a flow function;
Windmax—the function of calculation of wind speed at a vortex core border;
Windshape1—the function of calculation of longitudinal wind shear;
Windshape2—the function of calculation of cross wind shear;
Windshape3—the function of calculation of the F-factor (the index of danger of wind shear);
Windvelocity—the function of calculation of radial velocity of wind in the BCS.

Annex 5

Geometrical Ratios at Incomplete Reflection of Radar Signals from a Meteoobject

Let us consider a problem of assessment of EER of the allowed MO volume at incomplete reflection of the probing signals (the Fig. A5.1). At the same time a part of the allowed volume does not contain HM and does not participate in formation of the reflected signals.

In the simplest case presented in Fig. A5.1 the part is cut from the allowed volume of a cylindrical form (in a distant zone) by the horizontal plane (the upper MO bound) at the angle β.

Then

$$\begin{aligned}
|AK| &= z_{\max}/\cos\beta, \\
|BK| &= |BO| + |OF| + |FK| = r\Delta\beta + r\,\mathrm{tg}\,\beta + z_S/\cos\beta, \\
|AB| &= |BK| - |AK| = r\Delta\beta + r\,\mathrm{tg}\,\beta + z_S/\cos\beta - z_{\max}/\cos\beta \\
&= r(\Delta\beta + \mathrm{tg}\,\beta) - (z_{max} - z_S)/\cos\beta.
\end{aligned}$$

Let us estimate the value of the allowed volume filled with HM. According to the theorem of a body volume calculation on the areas of its parallel sections [19] (the Fig. A5.2),

$$V = \int_a^b S(x)dx. \tag{A5.1}$$

Let us introduce the system of coordinates as it is shown in Fig. A5.3 and consider the section of this cylinder (the allowed volume) by the planes perpendicular to the Ox axis. Let's calculate the area of section by the plane passing through the point M with the abscissa x, $|x| \le r\Delta\beta$.

Vereshchagin A. V. et al., *Signal Processing of Airborne Radar Stations*, Springer Aerospace Technology, https://doi.org/10.1007/978-981-13-9988-6

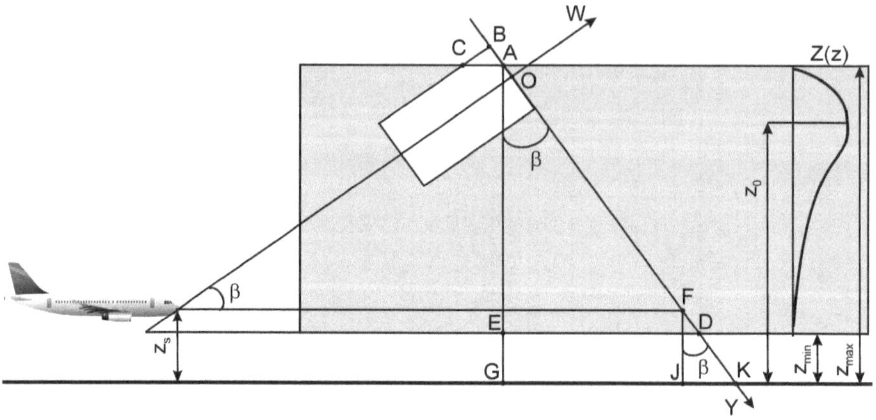

Fig. A5.1 Spatial ratios at incomplete reflection of signals from the allowed volume

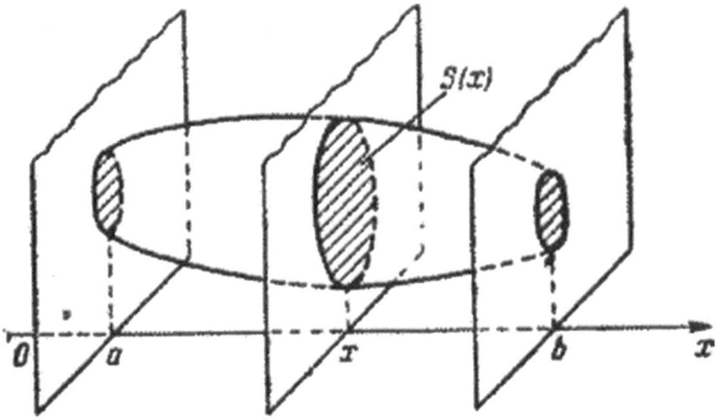

Fig. A5.2 Body volume assessment through the areas of its cross sections [19]

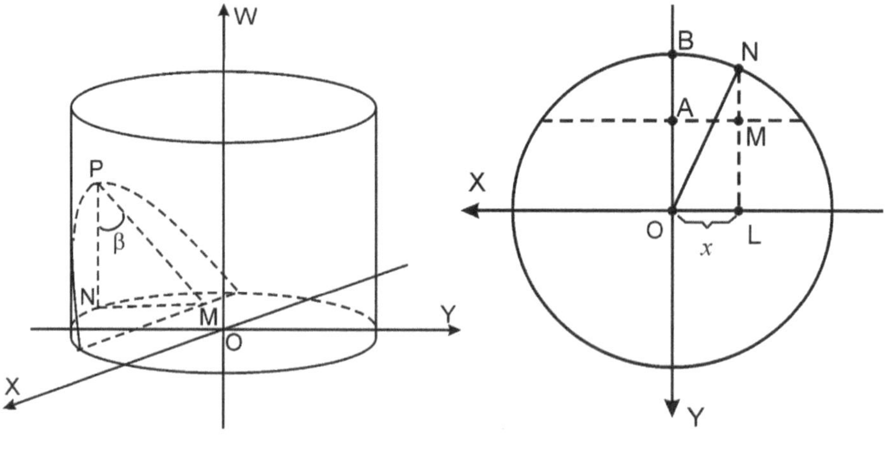

Fig. A5.3 Section of the allowed volume by the plane corresponding to upper MO bound

This section represents a rectangular triangle MNP therefore its area is

$$
\begin{aligned}
S_{MNP} &= \frac{1}{2}|MN||NP| = \frac{1}{2}|MN|^2\mathrm{tg}\beta = \frac{1}{2}(|NL| - |ML|)^2\mathrm{tg}\beta \\
&= \frac{1}{2}\left(\sqrt{|ON|^2 - x^2} - [|OB| - |AB|]\right)^2\mathrm{tg}\beta \\
&= \frac{1}{2}\left(\sqrt{(r\Delta\beta)^2 - x^2} - [r\Delta\beta - (r(\Delta\beta + \mathrm{tg}\,\beta) - (z_{\max} - z_S)/\cos\beta)]\right)^2\mathrm{tg}\beta \\
&= \frac{1}{2}\left(\sqrt{(r\Delta\beta)^2 - x^2} + [(r\,\mathrm{tg}\,\beta - (z_{\max} - z_S)/\cos\beta)]\right)^2\mathrm{tg}\beta.
\end{aligned}
$$

Integration limits in (A5.1):

$$
\begin{aligned}
a, b &= \pm\sqrt{(r\Delta\beta)^2 - |AO|^2} = \pm\sqrt{(r\Delta\beta)^2 - [r\Delta\beta - (r(\Delta\beta + \mathrm{tg}\,\beta) - (z_{\max} - z_S)/\cos\beta)]^2} \\
&= \pm\sqrt{(r\Delta\beta)^2 - [r\mathrm{tg}\,\beta - (z_{\max} - z_S)/\cos\beta]^2}
\end{aligned}
$$

At calculation of EER it is necessary to weigh in addition the vertical sections of the allowed volume (or its cut part) by the distribution of RR (water content) on the corresponding coordinate.

Then

$$
\sigma = \sigma_{no} - \int_a^b S(x)\sigma_d(x)dx, \tag{A5.1}
$$

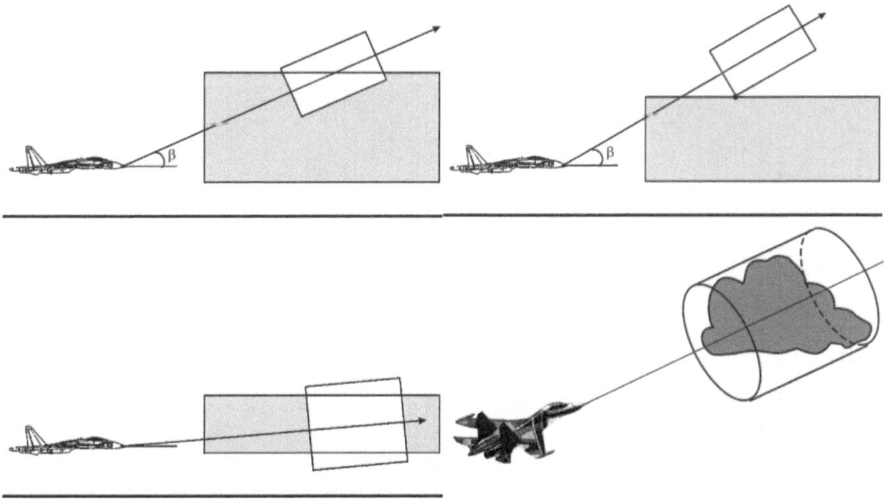

Fig. A5.4 Variants of spatial position of AV and MO at incomplete reflection of signals from the allowed volume

Here σ_{no}—the value of EER of the allowed volume calculated in an assumption of full reflection of the probing signals.

The value of the signal reflected by HM within the allowed volume can be estimated on computer by integration by numerical methods with use of corresponding software.

Besides the considered simplest case of incomplete reflection (filling of HM) of the MO allowed volume other spatial cases (Fig. A5.4) are also possible.

Bibliography

1. Adaptive spatial doppler processing of echo signals in RS of air traffic control. G.N. Gromov, Yu.V. Ivanov, T. G. Savelyev, E.A. Sinitsyn. SPb.: FSUE VNIIRA, 2002. 270 pages.
2. Active and passive radar-location of the storm and thunder centres in clouds/ Eds. L.G. Kachurin and L.I. Divinsky. SPb.: Hydrometeoizdat, 1992. 216 pages.
3. Beliy Yu.I., Zagorodny V. NIIP and its radars Bulletin of aircraft and astronautics, 2002, No., pages 107-109.
4. Vereshchagin A.V., Mikhaylutsa K.T., Chernyshov E.E. Improvement of structure and methods of processing of signals in multipurpose airborne radar stations for information support of increase in safety of flights in difficult meteoconditions: Theses of the report. In the book: Scientific and practical conference “Multipurpose Radio-electronic Complexes of Perspective Aircraft”: Theses of reports. SPb: HK “Leninets”, 2001. p. 1–14.
5. Vityazev V.V. Digital frequency selection of signals. M.: Radio and Communication, 1993. 239 pages.
6. Volkoyedov A.P. Radar equipment of planes: Ed. book. M.: Mechanical engineering, 1984. 152 pages.
7. Vorobiev V.G., Zyl V. P., Kuznetsov S.V. Complexes of digital flight navigation equipment. P. 1. Complex of standard digital flight navigation equipment of the Il-96-300 plane. M.: MGTU GA, 1998 140 pages.
8. Gorelik A.G. Radar methods of research of turbulence of the atmosphere. M.: Hydrometeoizdat, 1965. 25 pages.
9. Zubkovich S. G. Statistical characteristics of the radio signals reflected from the land surface. M.: Sov. Radio, 1968. 224 pages.

Vereshchagin A. V. et al., *Signal Processing of Airborne Radar Stations*,
Springer Aerospace Technology, https://doi.org/10.1007/978-981-13-9988-6

10. Kulemin G.P. Radar inferences from sea and land of radar station of centimetric and millimetric ranges. In the book: International scientific and technical conference "Modern Radar-location": Scientific and technical collection (Reports. Issue No. 1). Kiev: Scientific Research Institute Kvant, 1994. p.23–29.
11. Sharonov A.Yu. Influence of weather conditions on flight operation safety according to ICAO. In the book: Influence of the external environment on flight operation safety and issues of interaction of local factors on its condition: Interuniversity thematic collection of scientific works. L.: OLAGA publishing house, 1985. p. 89–91.
12. Sharonov A.Yu. Weather conditions at the time of plane crashes in ICAO member countries. In the book: Meteorological phenomena dangerous to flights and flight operation safety: Interuniversity thematic collection of scientific works. L.: OLAGA publishing house, 1984. p. 82–86.

References

1. Ivan M. A ring-vortex downburst model for flight simulations. Journal of Aircraft, 1986, v. 23, [1]3, p. 232–236
2. Ryzhkov A.V. Influence of inertia of hydrometeors on statistical characteristics of a radar signal. Works of GGO, 1982, issue 451, pages 49–54
3. Analog and digital filters: Ed. book S.S. Alekseenko, A.V. Vereshchagin, Yu.V. Ivanov, O. V. Sveshnikov; Eds. Yu.V. Ivanov. SPb.: BGTU VOENMECH, 1997 140 pages
4. Ivanov, Yu.V. Algorithms of space-time processing of signals in radio engineering systems: Ed. book. L.: LMI, 1991 101 pages
5. Marple Jr. S.L. Digital spectral analysis and its applications: The translation from English M.: Mir, 1990. 584 pages
6. Gong S., Rao D., Arun K. Spectral analysis: from usual methods to methods with high resolution. In the book: Superbig integrated circuits and modern processing of signals Eds. Gong S., Whitehouse H., Kaylat T.; the translation from English eds. V.A. Leksachenko. M.: Radio and Communication, 1989. P. 45–64
7. Khaykin S., Carry B.U., Kessler S.B. The spectral analysis of the radar disturbing reflections by the method of the maximum entropy. TIIER, 1982, v. 70, No. 9, pages 51–62
8. Mikhaylutsa K.T., Ushakov V.N., Chernyshov E.E. Processors of signals of aerospace radio systems. SPb.: JSC Radioavionika, 1997. 207 pages
9. Battan L. G. Radar meteorology. Translation from English L.: Hydrometeoizdat, 1962. 196 pages
10. Melnikov V.M. Connection of average frequency of maxima of an output signal of the doppler radar with characteristics of the movement of lenses. Works of VGI, 1982, issue 51, p. 17–29
11. Mironov M.A. Assessment of parameters of the model of autoregression and moving average on experimental data. Radio engineering, 2001, No. 10, pages 8–12
12. Bakulev P.A., Koshelev V.I., Andreyev V.G. Optimization of ARMA-modelling of echo signals. Radio electronics, 1994, No. 9, pages 3–8. (News of Higher Educational Institutions)
13. Bakulev P.A., Stepin V.M. Features of processing of signals in modern view RS: Review. Radio electronics, 1986, No. 4, pages 4–20. (News of Higher Educational Institutions)
14. Vostrenkov V.M., Ivanov A.A., Pinskiy M.B. Application of methods of adaptive filtration in the doppler meteorological radar-location. Meteorology and hydrology, 1989, No. 10, pages 114–119
15. Hovanova N.A., Hovanov I.A. The methods of the time series analysis. Saratov: Publishing house of GosUNTs "College", 2001. 120 pages

Vereshchagin A. V. et al., *Signal Processing of Airborne Radar Stations*,
Springer Aerospace Technology, https://doi.org/10.1007/978-981-13-9988-6

16. Andreyev V.G., Koshelev V.I., Loginov S.N. Algorithms and means of the spectral analysis of signals with a big dynamic range. Radio electronics issues Ser. RLT. 2002, issue 1–2, pages 77–89
17. Abramovich Yu.I., Spencer N.K., Gorokhov A.Yu. Allocation of independent sources of radiation in not equidistant antenna arrays. Foreign radio electronics. Achievements of modern radio electronics, 2001, No. 12, page 3–17
18. Yermolaev V.T., Maltsev A.A., Rodyushkin K.V. Statistical characteristics of AIC, MDL criteria in a problem of detection of multidimensional signals in case of short selection: Report. In the book: The third International conference "Digital Processing of Signals and Its Application": Reports. M.: NTORES, 2000 Volume 1. P. 102–105
19. Vilenkin N.Ya., Kunitskaya E.S., Mordkovich A.G. Mathematical analysis. Integral calculus. M.: Prosveshcheniye, 1979, 176 pages